False

False

How Mistrust, Disinformation, and Motivated Reasoning Make Us Believe Things That Aren't True

Joe Pierre, MD

OXFORD
UNIVERSITY PRESS

Oxford University Press is a department of the University of Oxford. It furthers the University's objective of excellence in research, scholarship, and education by publishing worldwide. Oxford is a registered trade mark of Oxford University Press in the UK and certain other countries.

Published in the United States of America by Oxford University Press
198 Madison Avenue, New York, NY 10016, United States of America.

© Joe Pierre 2025

All rights reserved. No part of this publication may be reproduced, stored in a retrieval system, or transmitted, in any form or by any means, without the prior permission in writing of Oxford University Press, or as expressly permitted by law, by license, or under terms agreed with the appropriate reproduction rights organization. Inquiries concerning reproduction outside the scope of the above should be sent to the Rights Department, Oxford University Press, at the address above.

You must not circulate this work in any other form
and you must impose this same condition on any acquirer.

Library of Congress Cataloging-in-Publication Data
Names: Pierre, Joe, author.
Title: False : how mistrust, disinformation, and motivated reasoning make us believe things that aren't true / Joe Pierre.
Description: New York, NY : Oxford University Press, [2025] | Includes bibliographical references.
Identifiers: LCCN 2024039799 | ISBN 9780197765272 (hardback) | ISBN 9780197765302 (ebook other) | ISBN 9780197765296 (epub) | ISBN 9780197765289
Subjects: LCSH: Delusions. | Belief and doubt. | Misinformation—Psychological aspects. | Disinformation—Psychological aspects.
Classification: LCC RC553.D35 P54 2024 | DDC 616.89—dc23/eng/20241029
LC record available at https://lccn.loc.gov/2024039799

DOI: 10.1093/oso/9780197765272.001.0001

Printed by Sheridan Books, Inc., United States of America

To my son, that he may have a better tomorrow.

Contents

Preface — ix
Acknowledgments — xv

1. Delusions, Distortions, and Misbeliefs, Oh My! — 1
2. The Psychology of Overconfidence — 13
3. Confirmation Bias on Steroids — 27
4. The Flea Market of Opinion — 42
5. The Disinformation Industrial Complex — 62
6. Conspiracy Theories Gone Wild — 86
7. Falling for Bullshit — 111
8. Divided States — 133
9. We Are Not Our Beliefs — 155
10. A Prescription for a Post-Truth World — 177

References — 193
Index — 229

Preface

Whether you believe the human brain is a divine creation sculpted spontaneously by the hand of God or the result of evolutionary forces converging over millennia, there's little doubt that it deserves to be recognized as a truly remarkable feat of engineering. Somehow, through processes that psychologists and neuroscientists have barely begun to fully understand, the mental faculties that have come to define our humanity such that we take them for granted—awareness, perception, memory, reasoning, imagination, abstraction, emotions, language, and consciousness—emerge from the three-pound, squishy mass of bioelectrically interconnected neurons that is the brain.

It's our brains that most set us apart from the other living beings on the planet and our brains that we can thank for our crowning achievements in art, design, construction, technology, and scientific discovery. Despite all the romantic mythology that surrounds the heart, we can survive on life support when it stops and can still come out of the operating room the same person after receiving a cardiac transplant. But when our brains die, we die. And even if we were somehow able to undergo a brain transplant, the essence of our individual identities would no doubt cease to exist. In that sense, we are our brains. And if human beings are remarkable, it's largely because our brains are remarkable.

But while the human brain is well-deserving of wonderment and perhaps even pride, this book is no celebration of the cerebrum. And while including them might improve its chances of becoming a bestseller, it doesn't offer any inspirational but false claims about the brain's untapped potential. It's not true, for example, that we only use 10% of our brains, as if that might mean we're capable of superhuman abilities like extrasensory perception, mind-reading, or telekinesis. The brain may be a veritable biological supercomputer, but it's not all that. Instead, what this book provides is a darker account and cautionary tale about how, for all the amazing and self-defining attributes of the human brain, many of its processes have the potential to wreak havoc in our daily lives and undermine our optimal social functioning.

Right now, there's no timelier and more worrisome example of cognitive processes that can thwart our ability to get along with each other than those that govern how we come to believe. Brains, belief, and behavior are inextricably linked so that we experience one leading to the next within a chain of causation. But while this higher-order ability to think and act allows us to transcend our animal instincts in favor of uniquely human reasoning, the result isn't always as adaptive as we might hope it to be.

Indeed, we're now living at a time when false beliefs are running rampant. And not only false beliefs in the form of superstitions, beliefs in the paranormal, or religious dogma that have always been embedded in the human psyche, but misbeliefs about terrestrial matters of the day that are both falsifiable and eminently consequential so that we end up behaving in ways that don't serve our best interests and can sometimes be downright harmful. It's gotten to the point that many of us no longer agree about how to decide whether or not something is true or what constitutes a fact. When factual evidence is presented to counter misinformation, it's dismissed as "fake news." Arguing, fighting, and even physical conflict have replaced legitimate debate that seeks to resolve disagreement. We've come to value "standing up for what they believe in" over considering alternative points of view and modifying our positions in light of new information. It's therefore been convincingly argued that we now live in a "post-truth" world—a world where we've become increasingly vulnerable to misinformation, and topics of scientific inquiry have been hijacked and transformed into political disputes, leaving countries, societies, and families around the world split apart by what we believe or don't believe.

This isn't only a matter of Thanksgiving Day dinner no longer being a safe space for friends and loved ones to gather in peace. Across the world, ideological polarization is tearing apart the very fabric of democracy and paving the way to civil war and authoritarianism. In this new era, with climate change looming over our heads in the wake of a global pandemic that claimed nearly 7 million lives over the course of just a few years, we're literally dying of suicide by false belief.

More than a decade ago, I attended a lecture where the speaker—a computer scientist from Cal Tech as I recall—claimed that the greatest threat to humanity was personal dogma. As a psychiatrist, I found this claim as intriguing as it was provocative, but the speaker seemed out of his lane, so I don't think I quite bought it at the time. But here we are at the start of 2025 and the evidence is all around us that he was right after all. So much of today's dysfunction can be traced back to our personal—and often false—beliefs. And our false beliefs, in turn, can be traced back to the smoking gun that is our brains.

* * *

To provide some perspective on the seemingly contradictory claim that our brains are at once extraordinarily capable and problematically flawed, I'll answer one of the most frequent questions I'm asked as a psychiatrist: "What's the difference between a psychiatrist and a psychologist?"[1] The simplest and most practical distinction is that psychiatrists go to medical school, are medical doctors (i.e., have an MD), and can prescribe medications, whether for depression or diabetes. Psychologists typically hold doctorate degrees (e.g., a PhD or PsyD) and are therefore correctly referred to as "doctors" (the word "doctor" is derived from the Latin *docēre*, meaning "to teach"), but aren't physicians and don't typically prescribe medications.

Beyond that, the difference can admittedly get confusing since there's considerable overlap between what psychiatrists and clinical psychologists often do—both diagnose mental illness, both use psychotherapy or "talk therapy" to treat patients spanning the continuum from the so-called worried well to those with more severe psychiatric disorders, and sometimes we work side by side as part of a collaborative team. In some jurisdictions, psychologists have even been granted prescribing privileges after completing additional medical and pharmacological training.

The real difference between psychiatrists and psychologists is less about what we do in clinical practice and more about the scope of psychiatry and psychology as academic disciplines and fields of study. Psychiatry, in its modern incarnation, is a medical subspecialty that focuses on psychopathology, including the diagnosis and treatment of people with mental disorders like schizophrenia, major depression, anxiety disorders, and dementia. Psychology is a largely scientific discipline with a much broader scope that extends beyond the clinical realm in its aim to understand all mental phenomena—how we think, reason, feel, and behave—whether normal or pathological.

Both psychiatry and psychology can tell us a lot about human beings and their behavior. In my application essay for psychiatry residency training programs, having completed an undergraduate degree at the Massachusetts Institute of Technology (MIT) studying both molecular biology and psychology and then medical school at the University of California Los Angeles, I wrote that "knowledge of normal human processes is often best obtained through the study of disease states." After all, I reasoned, our understanding of modern genetics largely emerged from research on genetic mutation. In much the same way, understanding mental illness and psychiatric symptoms like phobias and paranoia can inform an understanding of normal but potentially problematic phenomena like situational anxiety and fear.

Throughout my career as a psychiatrist, my main clinical interest has been working with people suffering from severe mental illness and especially psychotic disorders like schizophrenia. I've been particularly fascinated by the phenomenology of psychotic symptoms like delusions (idiosyncratic and false beliefs held with extraordinarily high conviction, like the belief that one is God) and hallucinations (perceptual experiences in the absence of external sensory stimuli, like "hearing voices" or "seeing things"). But while psychosis was the focus of my early career in clinical research and has been a long-standing topic of my teaching through the years, my academic work has just as often examined the intersection of psychiatry and psychology—that is, the gray area between psychopathology and normality where people aren't obviously mentally ill but aren't necessarily mentally healthy either. So, while I'm interested in delusions, I'm just as interested—if not more so these days—in less pathological "delusion-like beliefs" that include things like conspiracy theories.

Now you may be thinking that it's little wonder that a psychiatrist like me would serve up such a pathological view of the human psyche. After all, everyone assumes that "shrinks" are constantly analyzing people to find their faults or that we think everyone is "crazy," right?[2] But while psychiatrists may indeed be primed to sniff out evidence of mental disorder, I want to be clear from the outset that this book isn't about psychopathology per se. For the most part, it steers clear of discussing how the brain goes awry in psychiatric disorders and mental illness, whether in response to environmental insults, internal derangements, or the combinatorial interplay of the two. On the contrary, this is a story about normal people like you and me with normal brains that are simultaneously amazing but suboptimally functioning so that they have the potential to cause us significant problems.

In 1901, Sigmund Freud published a seminal contribution to psychoanalytic theory, *The Psychopathology of Everyday Life*, that postulated how unconscious processes lead to familiar and routine blunders like forgetting people's names or "slips of the tongue" (what we now call "Freudian slips").[3] In a similar way, *False* aspires to illuminate "the psychiatry of everyday life," drawing from the hybrid perspective of psychiatry and psychology to account for human belief and behavior. Chapter by chapter, it will unravel the psychology of false belief, beginning with an illustration of the borders of pathological delusion, a quantitative approach to understanding normal belief along a continuum of conviction, and a discussion of overconfidence and "positive illusions" in Chapters 1 and 2. Chapters 3, 4, and 5 focus on the roles of cognitive bias, motivated reasoning, and the illusory truth effect, respectively, in navigating the unprecedentedly vast landscape of information, misinformation, and disinformation where we search for knowledge. Chapters 6, 7, and 8, respectively, then tackle the timely and thorny topics of conspiracy theory belief, bullshit receptivity and pseudoscience, and political polarization that often lie at the heart of our ideological conflicts. Finally, Chapters 9 and 10 remind us that we are not our beliefs while offering a remedy to the pitfall of false belief through the "Holy Trinity of Truth Detection." Throughout its pages, *False* aims to elucidate how our brains interact with the world we live in today to make it all too easy to hold beliefs that put us at odds with one another and to passionately defend ideologies—sometimes to the point of violence—that often aren't worth defending. My hope is that reading this book will result in a kind of humbling self-awareness of our brains, our beliefs, and ourselves as less than perfect and that, armed with that insight, we might soften our convictions, view our ideologic opponents with compassion and understanding, and mend the rifts in our relationships as individuals and societies alike.

* * *

A half century before Freud, the journalist and writer Charles Mackay published his now classic work, *Memoirs of Extraordinary Popular Delusions* (later retitled *Extraordinary Popular Delusions and The Madness of Crowds*), a provocative telling of various historical infatuations and crazes that have swept over society, such as tulip mania, alchemy, fortune telling, and witch hunts. In it, he wrote,

> We find that whole communities suddenly fix their minds upon one object, and go mad in its pursuit; that millions of people become simultaneously impressed with one delusion, and run after it, till their attention is caught by some new folly more captivating than the first.[4]

Mackay's view suggests that there's a little psychiatric disorder in all of us and that the line that separates mental health and mental illness is fluid, such that we may find ourselves crossing over and back throughout our lives. This premise is echoed more concisely in the concluding voice-over narrative of the Hollywood movie adaptation of Susanna Kaysen's memoir *Girl, Interrupted*:

> Crazy isn't being broken or swallowing a dark secret. It's you, or me … amplified.[5]

Finally, a particularly blunt illustration of this claim can be found in a reference text on "acceptance and commitment therapy," a form of talk therapy that encourages us to be more willing to accept and weather difficult emotions in the name of positive change.

> Virtually all of our measures of "psychopathology" are built on the assumption that to be psychologically healthy is to be free of disordered emotional and cognitive responses. According to this standard, a coma victim might be considered the ideal of psychological health.[6]

Collectively, these quotations highlight the long-standing observation that the boundary between mental health and mental illness is often blurred, and, indeed, the idea that the distinction is more quantitative than qualitative is increasingly embraced in psychiatry. That doesn't mean that there's *no* meaningful distinction, or that the distinction is merely arbitrary, or that mental illness doesn't exist, as some have claimed. It means that the difference between mental health and mental illness can be understood in a way that's similar to how we understand colors within the visible light spectrum. Just as the distinction between "red" and "green" represents a quantifiable difference in wavelength, the difference between having a mental illness like schizophrenia or major depression and not having one can be understood as a matter of degree. But there's also a meaningful qualitative difference between mental illness and mental health, just as acknowledging a categorical distinction between red and green is crucially important when deciding what to do at a traffic light.[7]

And so, just as we shouldn't write off those who suffer from mental illness as some kind of "others" who have lost their humanity, we should acknowledge that being human and relatively mentally healthy doesn't preclude experiencing individual symptoms of mental illness that fall short of a full-blown disorder but might at some point become problematic. Anxiety, sadness, anger, impulsivity, forgetfulness, and delusion-like beliefs are part of a universal human experience shared by all of us at one time or another to various degrees. In fact, as I hope to make clear in this book, while the so-called mental status exam in psychiatry involves characterizing "thought processes" according to a normative default of thinking that is "linear, logical, and goal-directed," that hardly describes how most of us think all of the time. I'd go so far as to say that the idea that human beings always think rationally and logically, and form our beliefs accordingly, is a complete myth.

So it is that our propensity for false belief isn't a phenomenon of mental illness so much as it's a normal quirk of our imperfect brains. And it isn't a problem with only Republicans, or Democrats, or whomever it is that we identify as our ideological opposites, either. *False* isn't about "them" or "it," but about "us" and the world we live in.

Acknowledgments

This book would not have been possible without my wife's support and generosity with the invaluable gift of time. I'm also indebted to my agent, Bonnie Solow (and Jess Shatkin for connecting us), as well as my editor, Nadina Persaud at Oxford University Press, who both championed my work as a first-time book author. Thanks to Gordon Pennycook, John Jost, Paul Piff, Joseph Uscinski, and Adam Riess for discussing their research. And finally, a shout-out to my mother, who, through the years, kept asking so many of the questions I've tried to answer in this book.

1
Delusions, Distortions, and Misbeliefs, Oh My!

> Doubt is not a pleasant condition, but certainty is absurd.
> —Voltaire

Delusions

When I first met Vernon Roberts in the hospital, his face had the appearance of earthenware that had been fractured into shards, reassembled, and then glued back together, with the pieces not fitting together quite right. Thick scars that had long since healed formed a zig-zagging maze across a rugged dermal landscape. I guessed that Mr. Roberts, as a Vietnam War veteran, might have been the casualty of a landmine, but that didn't turn out to be the case. The answer to the puzzle of his face was something altogether different.

Mr. Roberts managed to survive the war with no major injuries to his external body, but, in the midst of his tour of duty as a 19-year-old, something else had started to eat away at him from within. As if being in constant fear for his life while fighting the North Vietnamese Army day after day wasn't enough, he began to feel increasingly like the members of his own platoon—the other young men who'd been fighting with him side by side for months—were plotting to kill him, too. Even more disturbing than that, he also developed the sense that his close friend Jimmy, one of his platoon mates who'd been killed in action some weeks before, had returned from the dead to take up residence in Mr. Roberts's body.

After being medevacked back to the states in 1971, Mr. Roberts was diagnosed with schizophrenia—a term originally coined a century ago to describe a kind of schism, or splitting, of the mind. He was started on antipsychotic medication that helped to reduce the intensity of his paranoia, but, through the years, he continued to believe that Jimmy—whom he would later come to refer to as "the pimp," "the puppeteer," and "the entity"—continued to reside inside him, with the ability to control his actions. This sense of lost control over oneself, which psychiatrists call a "somatic passivity experience" or "loss of agency," meant that sometimes when Mr. Roberts's body moved, it felt as if it was "the entity" inside him that was moving, not him.

To make matters worse, "the entity" often seemed to want to kill him. One evening, after he'd stopped his medications for several months, Mr. Roberts found himself standing atop a freeway overpass, looking down at an endless stream of cars speeding by underneath. To hear Mr. Roberts tell it, it was "the entity" that jumped, taking his body along for a freefall that ended abruptly when he smashed into the pavement and oncoming traffic. After a long hospital stay and extensive reconstructive surgery to repair his shattered skull and other bones, he defied the odds to survive the experience, just as he had during the war.

By the time I treated Mr. Roberts for a psychotic exacerbation many years later, he'd probably recounted his story dozens of times. And, each time he did, he always stuck to his claim that—contrary to how it read in his medical chart—the injuries he sustained from jumping from the freeway overpass weren't the result of a suicide attempt. He insisted that *he* never jumped; it was "the entity." One day however, he did stab himself in the stomach repeatedly with a knife, either in an attempt to rid himself of "the entity," or to end his life, or both. Once again, he lived to tell the tale.

"I'd rather kill myself than be killed," he explained.

Cases like Mr. Roberts have always fascinated me and sparked a curiosity about the possible mechanisms that might explain how delusional beliefs can take such a pernicious hold on one's psyche. But they've also made me do a lot of thinking through the years about just what exactly delusions are and how they can be reliably distinguished from other kinds of less pathological beliefs.

Picking up from the Preface, where I argued that there's a continuum as well as a gray area between normality and mental illness, this first chapter illustrates that false belief can also be conceptualized along a spectrum. Delusions, which psychiatrists consider to be prototypical symptoms of psychotic disorders like schizophrenia, lie at one end of that spectrum. At the other end are cognitive distortions that, while also treated by psychiatrists, offer an example of false belief so frequently encountered that virtually everyone has them. In keeping with the premise that they exist on a continuum, there are both qualitative and quantitative differences between these two phenomena.

* * *

In psychiatry, delusions are commonly defined as "fixed, false beliefs," where "fixity" is synonymous with incorrigibility or, as the fifth edition of the *Diagnostic and Statistical Manual of Mental Disorders* (DSM-5) puts it, "not [being] amenable to change in light of conflicting evidence."[1] The DSM-5 glossary adds a more formal definition as well:

> A false belief based on incorrect inference about external reality that is firmly held despite what almost everyone else believes and despite what constitutes incontrovertible and obvious proof or evidence to the contrary.

Like much of the DSM, this categorical definition has been critiqued and revised throughout the years and still remains debated. For example, it has been argued that delusions aren't only about external reality, and we could note that "incontrovertible" evidence is often unavailable to refute beliefs.

In practice, psychiatrists don't rigidly adhere to the DSM definition and often fall back on judging delusions much like everyone else does, as if they're readily apparent based on fixity and falsity in a "know it when I see it" kind of way. A psychiatrist's assessment is also informed by both clinical experience and context. For example, delusions are often recognizable due to stereotypical themes like persecution (e.g., paranoia about being followed or otherwise in danger), grandiosity or grandeur (e.g., beliefs that one is God or has "special powers"), or health (e.g., concerns about being infested by a parasite or having a monitoring device implanted in one's body). In addition, delusions found in disorders like schizophrenia are often accompanied by other symptoms, like hallucinations or "hearing voices." And yet, in practice, assessing the falsity of delusions based on their subjective strangeness or perceived impossibility—what psychiatrists have historically called "bizarreness"—is complicated by how hard it is to agree about what is or is not possible in the universe.[2] This is especially true for certain types of beliefs, like those of a religious or metaphysical nature, that are unfalsifiable. Does God exist, and does He talk to people? Does the "soul" persist after death? Is it possible that we're living in a computer simulation or a multiverse? Who can really say?

Beyond this falsity problem, another challenge of defining delusions in categorical terms is that the words we use to do so often fail, as Plato put it, to accurately "carve nature at its joints."[3] This problem isn't unique to delusions or psychiatry—it's one that plagues most attempts at categorical definition regardless of the topic at hand. For example, let's say a child asks us to define a bird. We might start by replying that a bird is a kind of animal with wings that can fly. But then the child asks, "What about bugs and bats?" so that we augment our definition to specify that a bird is an animal with wings that can fly and has feathers. And then, when the child brings up penguins, we modify our definition yet again to specify that a bird is an animal that has wings and feathers that usually can fly, but not always. And so on.

Many of us have lived through another real-life example of how taxonomies, or classification systems, have required ongoing modification to maintain definitional integrity. If you went to school before the 1990s, like I did, you spent half your life accepting that Pluto was one of nine planets in the solar system. But, in 1992, astronomers discovered other planetoids that were as big or larger than Pluto so that, forced with a decision to either add or subtract planets, Pluto was demoted to "dwarf planet" status. Now our children admonish us that there are only eight planets, not nine, and that Pluto isn't one of them.

In the same way, the quick and dirty definition of delusions as fixed, false beliefs fails to reliably distinguish them from seemingly similar phenomenon—like religious and political beliefs—that are often held with stubborn conviction in the face of conflicting or inadequate evidence. Following the DSM-5's guidance that delusions are "not ordinarily accepted by other members of the person's culture or subculture," psychiatrists often resolve this dilemma, as well as the problem of unfalsifiability, by substituting an assessment of a belief's idiosyncrasy—that is, whether or not it's shared by others. As a general rule, albeit with the rare exception in the form of shared psychotic disorder or *folie-á-deux* (French for "insanity of two"), delusions are not shared beliefs.

In my clinical experience, the unshareability of delusions hinges not on their apparent unfamiliarity or bizarreness, but on their self-referentiality.[4] In other words, delusions—like the stereotypical examples given earlier—are unshared because they're typically beliefs about the believer rather beliefs than about the world. For example, while it would be easy for you to find those who share the beliefs that there will be a Second Coming of Christ, that telekinesis is possible, that a microchip could be implanted in one's body for the purposes of tracking, or that alien abductions have occurred, it would be much harder to find those who agree that *you* are the Messiah, that *you* can move objects with your mind, that *you* have a microchip implanted in your brain, or that *you* were abducted by aliens.

Still, even if we were to conclude that delusions are fixed, false beliefs that are unshared and self-referential, categorical definitions only get us so far. At their boundaries, such definitions strain under pressure because they often represent attempts to establish discrete categories for phenomena that are in reality continuous. Where does black end and white begin when there are countless shades of gray in the gradient in between them? As I suggested in the Preface, one way around this dilemma is to specify the quantitative dimensions of a categorical entity. So, instead of defining the color "orange" based on what it looks like subjectively to our own eyes, we define it objectively based on wavelengths ranging from about 590 to 625 nanometers within the visible light spectrum. At the ends of that range, orange might appear more red or yellow to an observer, but if we measure its wavelength as falling within the definitional boundaries, we can accurately declare it to be orange nonetheless. In other words, "orange" isn't a singular entity but a quantifiable range that includes colors that are more or less orange.

Just so, the DSM acknowledges—as I also hinted in the Preface—that delusions can be thought of as a matter of degree, noting that "the distinction between a delusion and a strongly held idea is sometimes difficult to make and depends in part on the degree of conviction with which the belief is held despite clear or reasonable contradictory evidence regarding its veracity."[1] And yet, it offers little guidance on how to resolve that difficulty. Some psychologists and psychiatrists have therefore proposed that delusionality ought to be assessed by measuring "cognitive dimensions" that are quantifiable rather than categorical.[5] These include belief conviction (how much we believe something to be true), preoccupation (how much time we spend thinking about the belief), extension (the degree to which the belief pervades different aspects our lives), and distress (how much emotional unrest the belief causes). Such dimensions have less to do with a belief's content, or *what* is believed, and more to do with *how* the belief is held.

By thinking about delusionality in terms of these cognitive dimensions, it becomes possible to not only define delusions according to a quantifiable range but also predict how clinically relevant a delusion is—that is, how much disruption to one's functioning and well-being a belief is likely to cause so that it might warrant psychiatric intervention. Doing so also allows us to see why self-referentiality is so relevant to delusions—it often extends a belief beyond a claim about what's merely *possible* to a belief about what *is*. As noted earlier, many people share the belief that space aliens visit our planet and might even abduct people from time to time, but they aren't convinced that this is true and don't spend a lot of time thinking about it in any case. In other words, they hold the belief with low levels of conviction, preoccupation, extension, and distress. In contrast, an individual who has become convinced that *he* has been abducted by aliens so that he stays up all night with a loaded shotgun to prevent it from happening again holds the belief with high conviction, preoccupation, extension, and distress. That individual is much more likely to warrant the attention of a psychiatrist.

At the start of my academic career, when I was finishing my residency training in psychiatry, I wrote a paper attempting to think through the differences between pathological delusions and normal religious beliefs and to resolve them through a discussion of the quantifiable cognitive dimensions of belief. One of the paper's reviewers insisted that I offer a better definition of a delusion than the one in the DSM (DSM-IV at the time) that I was criticizing. This is what I came up with—it's not perfect, but I think it still holds up fairly well 20 years later:

> A delusion is a belief that is contradicted by objective evidence or is at variance with what most other people believe. Delusions are typically preoccupying, held with extreme and unassailable conviction, and influence one's behavior to a significant degree. Delusional thinking (like "anxiety") can span the continuum from normalcy to pathology, but is not alone indicative of mental illness or psychiatric disorder. Whether a delusion is an appropriate target for clinical attention depends on the context in which it occurs (i.e., along with other symptoms suggesting a particular disorder) and its impact on the individual or his or her relationship with others (whether it causes distress or dysfunction).[6]

Cognitive Distortions

Most of us are thankfully free of the kind of full-blown delusions that people with schizophrenia like Mr. Roberts suffer from. And yet we do often hold a variety of other "delusion-like beliefs" that can be fixed, false, preoccupying, and even self-referential, but are less idiosyncratic and fail to meet my definition of delusion due to lesser degrees of conviction and other cognitive dimensions of belief. In psychiatry and psychology, one of the most commonly encountered and insidiously troublesome examples of delusion-like beliefs are called "cognitive distortions."

Since the pioneering psychiatrist Aaron Beck introduced the term in the 1960s, cognitive distortions have been a foundational concept of a popular form of psychotherapy called *cognitive behavioral therapy* (CBT). While various definitions exist, cognitive distortions can be simply thought of as "errors of belief and how we arrive at and maintain them."[7] Though they fall short of delusions, Beck characterized them as "systematic deviations from realistic and logical thinking" representing "varying degrees of distortion of reality" that are "similar to [delusions] described in studies of schizophrenia.[8] During CBT, a help-seeking patient's cognitive distortions are identified and categorized as examples of all-or-none thinking, overgeneralization, jumping to conclusions, magnification and minimization, personalization, and the like.[9] The cognitive behavioral therapist then works with the patient to carefully examine the extent to which such beliefs are supported or refuted by objective evidence gathered by the patient during homework tasks completed outside of therapy sessions.

For example, let's say Edward is receiving CBT because he's feeling demoralized and anxious about how things are going at work. His therapist identifies potential cognitive distortions in his core beliefs that "everyone at work thinks I'm incompetent" and "I'm probably going to be fired." Initially, he estimates that he holds these beliefs with about 80% conviction or surety. During subsequent homework assignments during the week, Edward notes that while he still feels as if co-workers discount his input during meetings, his boss did mention that one of his proposals was a "good idea" and asked him to form a working group with several colleagues to pursue it further. He also reports that while he still believes he may be fired, his company has been downsizing due to economic woes and that other employees, not him, have been let go. In reviewing this counterevidence with his therapist, Edward agrees that he now only believes that "everyone" thinks he's incompetent with 50% conviction and that he only believes that he'll be fired with 40% conviction. In the following months, as he gathers more counterevidence, his conviction regarding his cognitive distortions diminishes further, along with how much time he spends ruminating about them. As a result, he reports feeling better.

This illustrative example highlights three important aspects of CBT that reveal something meaningful about cognitive distortions as well as beliefs more generally. First, CBT is based on the premise that cognitive distortions and beliefs lead to feelings and behaviors (e.g., "depression") rather than the other way around. Second, CBT demonstrates how beliefs can be modified by teaching patients to take a rational and more objective look at evidence that refutes them. And third, clinical improvement isn't necessarily judged based on the presence or absence of cognitive distortions but on quantifiable reductions in the cognitive dimensions of belief, such as the degree to which patients like Edward believe them and the extent to which the beliefs consume their thoughts.

Although CBT is often used to modify cognitive distortions in the context of psychiatric disorders like major depression, it has also been applied across the continuum of mental illness and mental health, with some success for delusions in schizophrenia as well as for cognitive distortions that aren't necessarily part of a mental disorder at all. In the above example, Edward felt "depressed" about work, but may not have met DSM-5 criteria for a full-blown major depressive episode due to still being able to enjoy his life outside of work and the absence of "neurovegetative" symptoms like insomnia or loss of appetite. Indeed, those of us "worried well" who aren't currently suffering from mental illness can often identify cognitive distortions that make us feel bad about ourselves from time to time. Not uncommonly, such distortions take the form of first- and second-person internal monologues or "self-talk" that includes beliefs like "I'm never going to get the job that I want," "You're too fat for other people to find you attractive," "I'm too lazy to ever write a book," or "You're never going to get married." While such self-referential beliefs often don't stand up to objective scrutiny, like when your friend tells you—and really believes—that you're not fat, they are nonetheless extremely common cognitive distortions among those who are self-critical. Regardless of whether such delusion-like beliefs represent symptoms of mental illness, CBT can help to talk people down from their cognitive distortions, exchanging them for more evidence-based beliefs.

Misbeliefs

As I suggested earlier, an inevitable challenge of defining terms like "delusion" is that the words used to define them often demand their own definitions. Just so, we could define "misbelief" as a broader category of erroneous belief that includes delusions, cognitive distortions, and a wide range of other false beliefs explored throughout this book, but we should first spend a moment attempting to define "belief" itself. While I've often searched for a satisfying definition, I have yet to find one.

The etymology of the word "belief" can be traced back to the Germanic word *lubh*, which means "to like or hold dear."[10] *Cambridge Dictionary* defines belief as "the feeling of being certain that something exists or is true,"[11] while the *Stanford Encyclopedia of Philosophy* calls it a "mental state, stance, take, or opinion about a proposition or about the potential state of affairs" and an "attitude we have, roughly, whenever we take something to be the case or regard it as true."[12] Two influential modern thinkers on the subject of belief and misbelief—Ryan McKay, a professor of psychology at the University of London, and the late philosopher Daniel Dennett—have written that a belief is "a functional state of an organism that implements or embodies that organism's endorsement of a particular state of affairs as actual."[13]

But what does any of that mean? What is a "state of an organism," an "attitude," or a "state of mind?" How is a belief different from a "stance," or an "opinion," or "knowledge?" And what does it mean for something to be "true?" Like the child who insists on asking "Why?" after every offered explanation, we could keep going on like this ad infinitum. I offer no solution to this dilemma—I like to define beliefs as "cognitive representations of past, present, and future reality, encompassing our inner experiences, the world around us, and the world beyond," but don't ask me to define "cognitive representation."[7]

Weighing the Evidence

Imperfect definitions aside, thinking about pathological delusions and not-so-pathological cognitive distortions reveals some key aspects of belief that set the stage for rest of this book. The first is that, as I mentioned in the Preface, it's a myth that human beings always think rationally. Thinking back to the DSM-5 definition of delusions, it's also a myth that human beings always hold beliefs based on carefully considering and weighing the evidence to support or refute them.

For example, paranoid delusions are common when under the influence of drugs like methamphetamines and often take the form of believing that one is being spied upon or in danger of being killed by one's neighbors. When I challenge patients with such concerns to consider whether their minds might be playing tricks on them or that their paranoia might be due to their drug use, they commonly answer with something to the effect of, "I *know* I'm not crazy, because it *seems* so real." In other words, as I mentioned earlier, delusional beliefs are typically defended based on anomalous subjective experiences—whether some kind of personal revelation or psychotic symptoms like hearing voices—amounting to a kind of "faith in subjectivity."[4] Indeed, recent research suggests that delusional thinking in schizophrenia may be at least partially explained by *circular inference*, in which direct experience is overweighted as evidence while information from others is underweighted.[14] Delusional thinking has also been attributed to a "jumping to conclusions" bias demonstrated when research subjects make up their minds prematurely without adequate evidence during probabilistic reasoning tasks.[15] Similar research using scales that measure quantitative cognitive dimensions of delusions have revealed that "delusion-proneness"—defined by subclinical or subthreshold levels of delusional thinking that fall short of frank psychosis—is common in the general population and is also associated with a greater degree of jumping to conclusions reasoning compared to those without delusion-proneness.[16]

Moving even further toward the healthy end of the mental illness–mental health continuum, a recent meta-analytic study (a study using a type of statistical analysis that combines the results of many similar studies looking at the same thing) tells us that, even for "normal" people without delusions or delusion-proneness, anecdotal evidence and first-person narrative accounts are often more persuasive than statistical evidence, at least under certain conditions.[17] As we might expect, this "anecdotal bias" is more likely to occur in situations involving high levels of emotional engagement and personal relevance, such as those related to health or threats. When emotional engagement is low, statistical evidence is more persuasive. Collectively, while also considering that narrative accounts do a much better job of capturing our attention and that most of the general population lacks a firm understanding of statistical mathematics, these research findings validate the notion that many of us are prone to rush to judgment and to prioritize anecdotal evidence over objective evidence, especially when forming emotionally charged beliefs. People with delusional beliefs or delusion-proneness may do this more than the rest of us, but this is a quantitative difference. We're all vulnerable to jumping to conclusions and anecdotal bias in varying degrees.

The distinction between subjective evidence and objective evidence is a crucial one that highlights just how loosely the term "evidence" is used when we defend our beliefs. Although the DSM-5 defines delusions based on "incorrect inferences" and conviction in the face of "evidence to the contrary," many of our nondelusional beliefs are based less on objective evidence than they are on intuitions, hunches, feelings, and other subjective impressions. For example, while some of us might decide whether to take a hit in blackjack based on the number of aces still in the deck or make a list of pros and cons when deciding if we should marry a potential spouse, most of us make such decisions based on our gut instincts. We believe that we'll get blackjack because we feel lucky, and we believe that we should marry someone because we're in love. And yet feeling lucky or in love often represents the lowest-quality evidence for accurately predicting outcomes, which helps to explain why the house tends to win in gambling and so many marriages end in divorce. Perhaps Freud said it best: "Ignorance is ignorance; no right to believe anything can be derived from it."[18]

In addition to intuition and gut instinct, the evidence to support our beliefs often also hinges on subjective experience, as illustrated by the popular dictums "seeing is believing" and "perception is reality." In psychology, the tendency to accept what we see as the "true reality" and to reject other people's accounts and objective counterevidence as faulty is known as "naïve realism." Lee Ross and Andrew Ward, the Stanford University psychologists who coined the term in 1996, described it as the faulty assertion that,

> I see entities as they are in objective reality, and that my social attitudes, beliefs, preferences, priorities, and the like follow from a relatively dispassionate, unbiased, and essentially "unmediated" apprehension of the information and evidence at hand.[19]

To say that naïve realism is a faulty assertion isn't to claim that personal or lived experience ought to be dismissed out of hand. On the contrary, what applies to others more generally doesn't always apply to ourselves in the moment, so that subjective experience may indeed be more vital under circumstances of emotional engagement, as when we're facing an imminent threat. But, as with intuitions and hunches, subjective experience represents evidence that can also be highly error prone, whether we're delusional or not. Adopting a "seeing is believing" approach risks mistaking hallucinations and mirages as real, eating poisonous berries because they look delicious, concluding that the Earth is flat, and taking for granted that human beings have free will merely because that's how it seems.[20] Some neuroscientists like Anil Seth at the University of Sussex have gone so far as to argue that normal perception is a kind of "controlled hallucination"—that is, a cognitive representation of reality analogous to beliefs but hardly a perfect reflection of it.[21] Although people these days often seem to exalt the idea of "personal truth," the more traditional concept of "truth" is based on the premise of a universal and objective reality, not one that's subjective.

* * *

The scientific method was designed to get us closer to objective and universal truth by weeding out subjective misinterpretations. During his 2016 commencement speech to Cal Tech, the physician and writer Atul Gawande described science as a "commitment to a systematic way of thinking, an allegiance to a way of building knowledge and explaining the universe through testing and factual observation."[22] This way of thinking is based on an iterative process of gathering evidence from repeated observations while controlling for different explanatory factors in order to inform theories about the true nature of the world and how things work. In other words, when we talk about scientific research, we're referring to "re-search"—a process of looking again and again to determine how much we can rely on what we observe.

But while objective scientific data represents some of the highest-quality evidence available, science has a long history of shortcomings that limit its mass appeal. First, although some people may be more naturally inclined to analytical thinking, the scientific method is hardly an intuitive way of forming beliefs. It's a skill that has to be learned and, in much the same way as CBT, requires that we retrain ourselves to prioritize objective evidence over our more subjective intuitions and experiences. Second, many believe that some questions lie outside the scope of scientific inquiry, such as whether the soul or God exists, citing the argument that "absence of evidence isn't evidence of absence." And third, although scientific discovery can be inspiring and exciting for scientists, nonscientists often regard science, rationality, and analytical thinking as buzz-kills that suck the life out of the party, leaving people feeling cold. In a world where only rational scientific thinking existed, delusions wouldn't exist but neither would religion, myths, magic, fiction, or fantasy. Human beings aren't computers or automatons—it would be unrealistic to expect us to tolerate a world so dull. And so it's easy to understand why we're living in a time when the pendulum of popular sentiment seems to be swinging away from valuing the kind of objective evidence that science relies on.

It's also worth acknowledging that no one—not even scientists—goes about their daily lives conducting controlled experiments to decide what to believe. Beyond our subjective intuitions, perceptions, and experiences, most of our beliefs are based on what others tell us, especially when we lack firsthand experience about a given subject. Much of what we learn as children from our parents and in school, as well as subsequently throughout our adult lives, requires a concession that we lack knowledge so that we must trust in informational sources that know better than we do based on their own experience or expertise. Such reliance on the testimony of others involves a leap of faith that's often based on our assessment of someone's character as much as on the content of the information in question.[23] Conversely, when people talk about mistrusting science and discounting expertise, they're often talking about mistrusting scientists and experts rather than science itself—as Gawande put it, "few dismiss the authority of science . . . they dismiss the authority of the scientific community."[22]

I'll say more about the role of mistrust in belief formation as it relates to conspiracy theories in Chapter 6. For now, I'll conclude by saying that while beliefs are based on different types of evidence, all too often the evidence is flimsy, with little popular agreement about what constitutes a hierarchy of evidentiary quality. No wonder, then, that people's individual beliefs are so often at odds with each other or just plain wrong.

The Paradox of Faith

That delusions and cognitive distortions can be modified through CBT highlights that beliefs can be thought of as not only cognitive representations but also as probability judgments. When we talk about belief conviction, we're speaking of how strongly or passionately we believe and how likely we think something to be true. We can quantify this—as we did in the earlier example of Edward—as a percentage, but paradoxically, the conviction we have for our beliefs is often *inversely* correlated with the objective evidence to support them. This paradox is well-captured by Robert Pirsig in *Zen and the Art of Motorcycle Maintenance*.

> You are never dedicated to something you have complete confidence in. No one is fanatically shouting that the sun is going to rise tomorrow. They *know* it's going to rise tomorrow. When people are fanatically dedicated to political or religious faiths or any other kinds of dogmas or goals, it's always because these dogmas or goals are in doubt.[24]

Scrutinizing that claim more closely, we might argue that most of us believe the sun is going to rise each morning with nearly 100% conviction. After all, clouds and eclipses aside, human beings have never witnessed otherwise. But when we speak of conviction in this way, we're talking about a kind of emotionless belief that doesn't require much willing or action on the part of believers. That's something different than the fervent passion of faith.

I'll return to the phenomenon of ideological passion in Chapter 9, but for now, suffice it to say that the distinction between mere belief and "dedication" to one's faith, as Pirsig put it, has been raised by a long tradition of thinkers throughout history. For example, the Bible says "faith is the substance of things hoped for, the evidence of things not seen,"[25] while the Renaissance philosopher Michel del Montaigne noted that "nothing is so firmly believed, as what we least know."[26] Friedrich Nietzsche offered a characteristically darker take: "'Faith' means not wanting to know what is true."[27]

The argument that faith warrants distinction from more pedestrian belief has also been made in the modern era. Daniel Dennett drew a line between "beliefs" and "opinions," noting that the latter term is preferred when describing probabilistic judgments or "bets on truths" that involve language, action, and commitment.[28] In a similar fashion, Oxford philosophy professor H. H. Price distinguished between "believing in" and "believing that," which John Byrne, author of the website *Skeptical Medicine*, has subsequently expanded on.[29] According to Price and Byrne, "believing that" something is true is a relatively straightforward matter of looking at the evidence, whether subjective or objective. In contrast, we "believe in" something when there's inadequate evidence to prove it or when the belief isn't falsifiable, as with religious or metaphysical beliefs. Finally, Joel Geiderman, a professor of emergency medicine at Cedars Sinai Hospital, provided a kind of synthesis of the seeming paradox of faith and uncertainty, noting that

> faith and doubt are two sides of the same coin. Faith is only possible when there is doubt. If something were truly known, without question, then believing it would not be an act of faith but rather an act of knowing. Faith is knowing in your heart what the mind cannot know.[30]

It might very well be useful to differentiate between beliefs based on objective evidence, on the one hand, and faith in the absence of objective evidence, on the other, with the latter representing not merely a feeling but an act of willing, similar to the distinction of being "in love" and "loving." Indeed, research suggests that the propensity to have that kind of faith exists in a kind of see-saw relationship with the propensity for analytical thinking, which could, for example, account for individual differences between theists and atheists.[31] Nonetheless, I'll steer clear of such distinctions in this book and leave discussions about the unfalsifiable beliefs of religion and metaphysics to the philosophers. Instead, I'll take a more practical approach going forward, modeling all beliefs as probability judgments while noting that many of our beliefs are held with excessive levels of conviction at the expense of acknowledging more appropriate levels of uncertainly. Stated another way, people often tend to adopt an all-or-nothing "belief that" attitude toward matters that warrant more probabilistic and opinionated "belief in." Such unwarranted conviction is the stuff of delusion and, as I'll argue throughout this book, often lies at the root of our ideological conflicts.

While it lies within our collective potential to uncover truths about the world we live in, we should understand that our beliefs are ultimately probability judgments and that few if any—including that the sun will rise tomorrow—deserve to be held with 100% conviction. And many of our beliefs should be held much less strongly. This isn't an argument against faith per se, but we should see faith for what it is: an active choice of believing in the face of uncertainty or lack of evidence. As Geiderman suggests, faith and doubt aren't mutually exclusive: for mentally healthy people, they can be maintained within a kind of integral yin-yang balance. We can have faith in things that might be unknowable while still working to improve our ability to think in a more rational and objectively evidence-based fashion and to assign more accurate probabilities to our beliefs. That might sound boring, but we know from experience with CBT that it's possible and that people can feel better when they embrace reality. When we instead choose to hold fast to delusion-like beliefs and self-deception, we do so at our peril.

2
The Psychology of Overconfidence

Ignorance more frequently begets confidence than does knowledge.
—Charles Darwin

Beliefs as Probability Judgments

A married couple I know had three beautiful daughters over the course of their thirties. Three kids always seemed like a good aspirational target for them, and, between their jobs, shuttling their kids to school, piano recitals, dance practice, birthday parties, and the like, they had their hands full. And so, as they entered their forties, they discussed whether it might be time to call it quits. But both of them had always wanted a son, and, after a long talk, they decided to give it one more try. When they got pregnant again after considerable effort, they were pretty sure they were finally going to have a boy, but when their obstetrician performed an ultrasound and offered the "gender reveal," it was—you guessed it—another girl.

I never asked my friends why they felt so confident they'd have a boy on their fourth go around. Like many beliefs that we hold, it was probably a combination of things. Maybe it was wishful thinking. Maybe it was just a feeling or the old wives' tale about "carrying low."[1] Or perhaps it was the idea that they already had three girls and were therefore due for a boy based on the "law of averages." If this last line of reasoning was the case, then my friends would have been guilty of a common cognitive error called the *gambler's fallacy*.

The most famous historical example of the gambler's fallacy is said to have taken place back in 1913, at the Monte Carlo Casino, in Las Vegas. During a game of roulette, the ball fell on black five times in a row, and then ten, and then twenty. As the streak went on, players increasingly placed their bets on red based on the belief that the law of averages meant that the streak of landing on black was increasingly likely to end. But that's not actually how the probabilities work—assuming the roulette wheel wasn't rigged, the chances of the ball landing on black were always 50-50 no matter the streak. In fact, it wasn't until the 27th spin of the Monte Carlo casino roulette wheel that the ball finally landed on red. By that time however, the gamblers who'd been upping their bets on red after each spin had lost millions of dollars based on the false expectation that they were increasingly sure to win.

At its core, the gambler's fallacy involves neglecting the statistical reality that chance events with a 50-50 probability are always just that, irrespective of prior sequences of events. That the gambler's fallacy is common would seem to suggest that many of us simply lack a solid foundation in basic statistical probability. That conclusion might not be unexpected since many people go through school without ever taking a formal course in statistics or data analysis.[2] But research in behavioral economics, a branch of economics that intersects with psychology, paints a more complicated picture that reveals that the gambler's fallacy and other statistical mistakes we make are also a matter of how our brains are sometimes led astray by faulty intuitions about probabilities. Unlike psychologists and psychiatrists, behavioral economists tend to focus less on personal beliefs like delusions and cognitive distortions and more on beliefs involving probability judgments and decision-making based on *Bayesian reasoning*, where the actual statistical probabilities and actuarial risks are more or less known—like in gambling or investing in the stock market.

In this chapter, I'll take a deeper dive into naïve realism and the paradox of faith by exploring the natural tendency we all have to not only be confident, but overconfident, about not only statistical probabilities but also much else that we believe. As we'll see, this tendency can be self-protective in healthy doses, but, as always, too much of a good thing can just as easily point us down a darker path.

***

In 1971, the psychologist and behavioral economist Amos Tversky and his colleague Daniel Kahneman—who would later go on to win a Nobel Prize after Tversky's death based on their collaborative work on "prospect theory"—reported that misleading intuitions like those that underlie the gambler's fallacy are present even among those who have had substantial training in statistics.[3] In other words, people who know better are still vulnerable to faulty intuitions that result in unwarranted confidence when making probability judgments.

Twenty years later, Tversky and his erstwhile graduate student Dale Griffin, now a professor at the University of British Columbia Sauder School of Business, performed experiments that further clarified that this kind of overconfidence when making faulty probability judgments sometimes occurs because people tend to overprioritize the "strength" of evidence at the expense of considering the "weight" of evidence.[4] To illustrate the difference between strength and weight, let's say we wanted to find out if the Monte Carlo Casino roulette wheel was rigged to preferentially come up black and therefore decided to monitor it closely over the course of a night. The proportion of times that we observed the ball landing on black would represent the strength of our evidence, whereas the weight would depend on the number of observations or repeated spins of the wheel. So, if we only observed 30 consecutive spins of the wheel and saw that the ball landed on black 26 times, like it did back in 1913, the strength of evidence suggesting that the wheel was rigged would be strong. But the weight of the evidence would depend on the total number of spins, so that if we observed the roulette wheel all night long and found that the ball landed on black 4,997 out of 10,002 spins, we'd more accurately conclude that the wheel wasn't rigged after all. We can see then that weight can put strength into better perspective.

Tversky, Griffin, and Kahneman's research tells us that whether or not we're well-versed in statistics, we're all vulnerable to the erroneous belief in the "law of small numbers" or an overreliance on the strength of small sample sizes at the expense of the bigger picture. We're most vulnerable to overconfidence when the strength of the evidence is high and the weight is low,[4] but our specific intuitions depend on a variety of other situational factors as well. Let's consider three examples. First, there's the gambler who falls victim to the classic gambler's fallacy described above. At the Monte Carlo Casino roulette wheel, he bets on red after 10 previous occasions that the ball landed on black because of confidence in the false belief that the "law of large numbers" will shift the 50-50 probability on future spins to make red more likely due to a kind of self-correction. Next, let's say another gambler won on his previous 10 bets on black. If he then puts all of his money on black again based on overconfidence that he's on a "lucky streak" associated with betting on black, he would be committing an error called the *reverse gambler's fallacy*. Finally, if space aliens visiting the casino, who didn't know anything about roulette and had no reason to know that there was a 50-50 chance of the ball landing on black or red, observed the ball coming to rest on black 10 times in a row, it would be perfectly rational for them to predict with confidence that the next spin would result in the ball landing on black yet again.

Note that, in all three scenarios, the gamblers' mistake isn't the bet itself—the probability of the ball landing on red or black on a single spin is always 50-50, so that even if it landed on black 26 times in a row, betting on either outcome on the 27th spin would represent equally good or bad decisions. Instead, the cognitive error relates to overconfidence in the expected outcome, the rationale underlying that confidence, and how that affected the amount of money placed on the bet. Since the space aliens were ignorant of the underlying "base rate" of roulette wheels coming up red or black on any given spin being 50-50, it's only the human gamblers' who are guilty of the *base rate fallacy* whereby that probability was neglected in favor of erroneous belief in the *law of small numbers* that lies at the heart of the gambler's fallacy and the reverse gambler's fallacy.[3,5]

It's cognitive errors like these, where the actual probability of events is known but misjudged by individuals in the moment, that seem to most interest behavioral economists. Accordingly, the lens of behavioral economics is often focused on trying to understand how our intuitions about probability match or mismatch with actual statistical probabilities and how we update our probability estimates based on new evidence. But, even within economics, such probabilities get much trickier once we move from betting on red or black on a roulette wheel to making decisions about more complicated outcomes, like trying to beat the stock market. And beyond economics, when we attempt to predict the kind of chaotic variables that govern human behavior, statistical probabilities become even less clear. After all, as the Holocaust survivor and fiction writer Ilona Karmel admonished when I was taking her creative writing course at MIT while pursuing a degree in molecular biology, "human bodies aren't machines." Still, as a general principle, what we can take to the bank from behavioral economics is that, even when probabilities are known and even when people are well-versed in statistical mathematics, "people are often more confident in their judgments than is warranted by the facts."[4]

The more specific tendency to overprioritize strength over the weight of evidence and selective personal observations over base rates, whether or not individuals are aware of the overarching probabilities, should sound familiar by now because it's essentially the same phenomenon reflected in the *jumping to conclusions* reasoning style and the overreliance on narrow subjective experience that's associated with delusional thinking and naïve realism that I mentioned in Chapter 1. Remember though that this is a quantitative difference such that although people with delusions and delusion-proneness might have more vulnerability to these kinds of errors, it's a universal vulnerability that we all share to some degree under certain conditions. In much the same way, the gambler's fallacy has been found to be overrepresented as a kind of cognitive error among "problem gamblers" but is still present in non–problem gamblers, and presumably the rest of us, as well.[6]

* * *

Following the Nobel Prize-winning work of Tversky and Kahneman, behavioral economics has provided us with a useful model to understand these kinds of errors in probability judgment as a result of intuitive cognitive shortcuts that fall under the larger umbrella of *heuristics*.[7] In his best-selling 2011 book, *Thinking, Fast and Slow*, Kahneman proposed two different modes or systems of decisional thinking: an automatic, fast judgment based on instinct, intuition, and emotion and a slower, more rational, and deliberative process.[8] Heuristics represent fast-mode thinking that ideally works together with more deliberative reasoning within a balanced, "dual process" harmony. However, errors in judgment can arise when one mode of thinking wins out over the other, just as when we fail to account for both the strength and weight of evidence. For example, thinking carefully and deliberately might be wise for some decisions, like how to invest our 401(k)s, but not for those where swift action is needed, like jumping out of the way of a car. Likewise, the impulsivity of quick, instinctive thinking could save us from getting hit by a car, but not if we failed to consider the additional risk of jumping out of the way of the car right into cross-traffic and the path of an even faster-moving bus. Although it would be best to integrate the strength and weight of evidence, base rates and personal observations, and fast and slow thinking to optimize decision-making, behavioral economics tells us that this often isn't how human brains work whether we're forming judgments in the first place or updating our beliefs over time.[9]

When fast-mode heuristics result in faulty cognitive representations of reality—that is, false beliefs—they're referred to as "cognitive biases." To date, it has been proposed that there are nearly 200 cognitive biases (including naïve realism, the gambler's fallacy, and the base rate fallacy) that have evolved over time to make human decision-making more efficient while also making us prone to errors in accurately judging the risks and benefits of our actions.[10] I'll be saying a lot more about other cognitive biases throughout the rest of this book, but one thing to keep repeating from the start is that they are universal biases that we all have to some degree and are not clearly related to cognitive ability or intelligence.[11] In other words, vulnerability to cognitive biases isn't about being "stupid" or "dumb." It's about being human.

The Healthy Lies We Tell Ourselves

Cognitive biases related to overconfidence, such as the base rate fallacy and the gambler's fallacy, can cause significant problems not just when gambling or investing, where we could lose the family fortune, but also in many other examples of decision-making based on assessments of risk and reward as well. Indeed, when world leaders decide how to best respond to a pandemic or whether to wage war, overconfidence in expected outcomes can prove disastrous, resulting in the loss of countless lives. Based on such potential for damage on a global scale, Kahneman has said that overconfidence is the cognitive bias he would most like to eliminate if he had a magic wand.[12]

As with delusions, delusion-like beliefs, and other continuous phenomena, the boundary between confidence and overconfidence is often blurry. Just as online dating profiles often seem to idealize a mate who's "confident, but not cocky," we all might aspire to a Goldilocksian ideal or sweet spot of "just right" or just enough confidence. What might be surprising is that some degree of overconfidence—that is, confidence that's unwarranted given the facts so that it amounts to false belief and self-deception—may actually bolster our mental health. That was the radical hypothesis of University of California Los Angeles psychologist Shelley Taylor back in 1988, when she, along with Jonathan Brown of Southern Methodist University, first outlined the case for "positive illusions" defined as misbeliefs associated with happiness, the ability to care for others, and the capacity for creative, productive work.[13] Put more simply, positive illusions are healthy lies that we tell ourselves.

Thirty years since Taylor's initial research, positive illusions have become well-recognized types of misbelief in psychology that fall into three general categories: the better than average effect, the illusion of control, and unrealistic optimism.

"I'm Better Than the Average Person"

In her original paper with Brown, Taylor cited evidence across many studies that we tend to regard positive traits as core parts of our identity while discounting the negatives. Most people also report that they're "better than the average person"—a mathematical contradiction if a trait is normally distributed within a Bell curve—with self-appraisals that are inflated compared to how we're regarded by others. This cognitive bias has come to be known as the "better than average effect," the "superiority illusion," or the "Lake Wobegon effect" after Garrison Keillor's fictional radio show community in which "all the women are strong, all the men are good-looking, and all the children are above average." Indeed, we tend to extend the superiority illusion beyond ourselves to our loved ones as well, offering an explanation for how love can be blind, such that we often overlook the faults and foibles of both our romantic interests and our children alike.

Subsequent research by Brown found that the better than average effect is stronger for valued attributes like honesty, kindness, responsibility, intelligence, and competence.[14] The effect also increases following threats to self-worth in the service of allowing us to still feel good about ourselves. But in contrast to such common and hard-to-measure personality characteristics, other research has shown evidence for a "worse than average effect," where we underestimate ourselves when it comes to less common abilities and difficult tasks, such as computer programming, riding a unicycle, or coping with the death of a loved one.[15] This suggests that the better than average effect may sometimes be less about self-aggrandizement than it is about misinterpreting the term "average" as a pejorative rather than a statistical norm.

Although the better than average effect has been found to be associated with psychological well-being, there's evidence that its benefits depend on quantity, in keeping with my suggestion of a Goldilocksian ideal. It should come as no surprise that confidence is correlated with self-esteem—after all, they're nearly the same thing. And overconfidence could very well lead to perseverance that's predictive of real achievement in some circumstances, such as among children learning new skills, or even superiority, such as among elite athletes. But it should also come as no surprise that the superiority illusion has also been correlated with narcissism, whereby "self-enhancing" individuals are more likely to be rated by others as condescending, resentful, and defensive.[16]

As always then, the devil seems to be in the details. "Self-enhancement," defined by a discrepancy between our self-perceptions and others' impressions of us, might have more negative effects than overconfidence that's not as obvious to others.[17] Some research has also suggested that self-enhancement might have short-term social benefits through favorable initial impressions that might end up being more negative and socially harmful in the long run.[18] Social psychologist Roy Baumeister has argued in favor of the Goldilocksian ideal for the better than average effect, describing an "optimal margin" whereby too much superiority illusion, but also too little, might be associated with less psychological well-being.[19] Just the right amount of overconfidence might therefore serve us well, though it would probably be best to keep our perceived superiority to ourselves.

"I Am the Master of My Fate"

Belief in personal control over circumstances that in reality lie beyond our control represents a second category of positive illusions. *Locus of control* is a well-known construct in psychology that refers to the more generic belief in how much personal control we have over life events, whether or not that belief is accurate. For example, those with a high degree of internal locus of control would be more likely attribute a good night at blackjack to their skill as a gambler than merely being the beneficiary of "the luck of the draw." Locus of control has been studied for well over 60 years, with research finding that belief in personal control as well as an exaggerated sense of personal control can have a variety of potential benefits.[20]

As with the better than average effect, however, there seems to be a healthy medium. By way of illustration, let's examine two quotations. The first comes from William Ernest Henley's poem *Invictus*. While the English poet was recovering from amputation-sparing surgery on his leg in 1870s, he wrote that despite his hardship:

I am the master of my fate, I am the captain of my soul.

This kind internal locus of control embodied by Henley's words has been shown to be associated with positive health outcomes for both physical illness and mental health alike.[21]

Now contrast that self-empowering spirit with the attitude expressed in a second quotation from Werner Herzog's 2005 documentary film *Grizzly Man* about the self-styled naturalist Timothy Treadwell, who, after many summers camping in the Alaskan wilderness to commune with the bears that he loved, was mauled to death and eaten along with his girlfriend. In his iconic dead-pan voice-over narrative, Herzog says:

I believe the common denominator of the universe is not harmony, but chaos, hostility, and murder.

Needless to say, although *Grizzly Man* is one of my favorite movies because of Treadwell's breathtaking footage and the character study Herzog provides through his film editing, Henley's quotation is far more likely to be featured on an inspirational poster or t-shirt. Indeed, the helplessness and hopelessness that we might appropriately feel as a result of Herzog's version of "depressive realism"[22] represent the antithesis of personal control and can often be found to lie at the root of psychiatric disorders like major depression and posttraumatic stress disorder. It's easy to see how illusions of control epitomized by Henley's famous verse might conversely buoy our spirits while attempting to navigate our lives in a cold and harsh world.

In keeping with a Goldilocksian ideal, however, illusions of control can clearly become more harmful when they're more obviously inaccurate and within certain settings, as when trying to befriend wild grizzly bears. But they also have the potential to exact a societal toll beyond individual foolhardiness. Some years ago, University of California Berkeley psychologist Paul Piff performed a well-publicized but as-of-yet unpublished experiment in which study participants played a rigged game of Monopoly that awarded disproportionate monetary advantages to some players at the start of the game and as it progressed.[23] At the game's end, the advantaged winners proudly attributed their success to personal skill and superior strategy rather than advantage or even luck. This, along with other research by Piff, suggests that illusions of control can result in unwarranted self-appraisals for people with inherent financial advantage, leading them to be less empathic toward those who are disadvantaged.[24] Such people might, for example, be more likely discount the ethical or practical benefits of real-world social programs like welfare or affirmative action. This conclusion suggests that while some illusions of control can result in higher self-ratings of individual happiness or mental health, they might also contribute to interpersonal disregard, with a harmful effect on society as a whole.

"The Future Will Be Great, Especially for Me"

The third category of positive illusions involves overestimating the likelihood that good things will happen to us while underestimating the risk of bad outcomes. This view of the future through rose-colored glasses has come to be known as the illusion of *unrealistic optimism* or simply *optimism bias*. Unrealistic optimism accounts for why many of us believe that we might defy the odds to have a long and happy marriage or win the Mega Millions lottery.

In the *Devil's Dictionary*, Ambrose Bierce defines an optimist as "a proponent of the doctrine that black is white" and a cynic as "a blackguard whose faulty vision sees things as they are, not as they ought to be."[25] Indeed, like illusions of control, unrealistic optimism is thought to represent a form of denial—the antithesis of depressive realism—that can reduce stress and anxiety and allow us to devote energy to achieving goals. Recognizing the overlap between unrealistic optimism and hope, we can appreciate how viewing the world as an optimist instead of a cynic might exert a kind of placebo effect on mental health that reflects more about how we feel about the world than how it actually is or will be. Once again, however, such excessive optimism can have a dark side, as when it results in a "planning fallacy" that can lead us to engage in dangerous behaviors like smoking, unprotected sex, or texting while driving, despite known risks.

Positive illusions are defined by their benefits on mental and physical well-being. When overconfidence becomes so excessive that it causes more harm than good, it's not a positive illusion anymore. And yet the cautionary tale that Herzog tells in *Grizzly Man* provides a sobering illustration of how difficult it can be to achieve the Goldilocksian ideal. On the one hand, Herzog uses the ample footage that Treadwell shot of himself to put his self-superiority, illusions of control, and unrealistic optimism on display in a way that makes it obvious that he was undone by "vaulting ambition" in the end. But, viewed from another perspective, Treadwell managed to survive 12 seasons in the Alaskan outback, largely on his own, before he fell victim to the bears with whom he was trying to commune. That was a remarkable achievement and one that garnered him some measure of notoriety, with an appearance on *The*

Late Show with David Letterman, in the years before his demise. The object lesson of Treadwell's life and death provides a stark demonstration that there's often a fine line between overconfidence that helps and self-deception that harms.

False Memories

Some years ago, I was reminiscing with my old friend CJ about a notorious escapade from our college days. It was a story that I knew well, having recounted it and heard it recounted among our circle of friends on many occasions. But this time was different because CJ and I disagreed on some key details. And when I tried to argue my version's accuracy, CJ reminded me that he had been there, after all, whereas contrary to how I remembered it, I had not. The revelation hit me, as they say, like a ton of bricks, because I could "see" the event in my mind, and myself in it, when I recollected the memory. But when I thought more about it, I realized that he was right. I hadn't been there—in fact, I'd been on the other side of the country at the time. This incident was one of the first times that I realized just how faulty my memory could be. And, as I've gotten older, similar instances with recollections of past events that are at odds with those of family and old friends have become increasingly frequent.

When my son was four years old, his memory put mine to shame. I barely remember anything before I was four or five, but he was able to recall events from when he was two or even younger. This seemed extraordinary to me until I considered that, from a very young age, he'd always enjoyed watching videos that I'd recorded on my phone since he was born—his first steps, his first words, his first snow, and the time he had to go to the hospital when he had a fever. Because he'd watched them so many times, I could never be sure if he remembered the actual events or if he just remembered what was in the videos. It was probably the videos, though I don't think that he was able to recognize the difference.

The good news for me is that my faulty memory is just fine. I'm not suffering from early-onset Alzheimer's disease or anything like that. The less encouraging news for all of us is that the frequency with which I'm unable to accurately recall certain past events represents a kind of universal cognitive deficit from which we all suffer. Most of us are well aware that we don't remember the majority of the things that happen in our lives and that many of our oldest memories are hazy and often less than accurate. But, over the course of her career, University of California Irvine psychologist Elizabeth Loftus has demonstrated that even our memories of recent events are highly susceptible to error, especially when we're cued to recall our memories in a specific way.[26] For example, in one of her first experiments from the 1970s, she showed films of a car accident to subjects and used different verbs to ask them how fast the car was going. When subjects were asked how fast the car was going when it "smashed" into the other car, they rated it going 10 miles per hour faster than when the verb "contacted" was used in place of "smashed." A week later, when asked whether they remembered seeing any broken glass in the film, subjects who'd been cued with the word "smashed" were more likely to say "yes." In another experiment, Loftus showed a film of a car failing to stop at a stop sign and getting into an accident after making a right turn. Half the viewers were then asked how fast the car was going when it ran the stop sign, while the other half were asked how fast the car was going when it turned right. Those cued with the word "stop sign" were significantly more likely to answer "yes" when asked if they had seen a stop sign in the film.

Decades of research by Loftus and other investigators have demonstrated robust evidence of this "misinformation effect," whereby memories can be manipulated by the way we're asked questions about past events, and not only in trivial ways. False memories involving various past events in one's life—like getting lost, being attacked by an animal, nearly drowning, and being in an accident—can be suggested to people who come to endorse them as reality.[27] As you might imagine, such findings have shaken the foundation of criminal justice, with the reliability of eyewitness accounts and testimonies in response to police interrogation called into question and notable cases of alleged childhood Satanic ritual abuse from the 1990s chalked up to leading questioning and the power of suggestion. Because Loftus's work—detailed in her 1994 book, *The Myth of Repressed Memory*—has shown that "repressed" or previously forgotten and "recovered" memories of experiences like sexual trauma can sometimes amount to "false memories,"[28] she has been called as an expert witness in the defense of accused rapists like Bill Cosby, Jerry Sandusky, and Harvey Weinstein.

To be clear, Loftus isn't necessarily claiming that those individuals were innocent, but rather that,

> If I've learned anything from these decades of working on these problems, it's this: just because someone tells you something with confidence, just because they say it with lots of detail, just because they express emotion when they say it, that doesn't mean that it really happened. We can't reliably distinguish true memories from false memories. We need independent corroboration. Such a discovery has made me more tolerant of the everyday memory mistakes that my friends and family members make.[29]

Loftus has been telling us this for years, but few have been listening or persuaded. A 2011 survey found that large portions of the public believe that memories are permanent, unalterable, and reliable, while a substantial minority believes that the testimony of a single confident eyewitness should be enough to convict a defendant of a crime.[30] Another survey by Loftus and her colleagues found that even clinical psychologists still endorse misbeliefs about the reliability of "repressed" memories.[31] But research evidence says they're wrong: our memories aren't really like video cameras as many imagine, storing faithful accounts of events that can be recalled at will the way my son does with my phone. In reality, our memories are often imperfect accounts of past events that are frequently edited and rewritten based on new information—and sometimes misinformation—to serve our needs in the present.[32] Therefore, like the paradox of faith that I discussed in the previous chapter, our confidence about the accuracy of our memories should in no way be mistaken for an indicator of reliability. While that may run counter to our subjective experience, the reality is that our memories are both fallible and malleable.

Of course, many of our memories are accurate, and some people's memories are more reliable than others. Vulnerability to the misinformation effect tends to be greatest in children and the elderly as well as in those with lower scores on measures of one's intelligence quotient (IQ).[33] But Loftus tells us that, despite such quantitative vulnerabilities, nobody is completely immune to the effect of misinformation on memory. And since many of our beliefs depend on our memories—and not only our beliefs about past occurrences but factual beliefs as well—we should be aware that when we use those memories as grounds to justify the confidence we have in our beliefs, the ground that we're standing on is shaky.

The Dunning-Kruger Effect

No discussion of overconfidence in the accuracy of one's judgments would be complete without covering a phenomenon that has become a household word in recent years, the *Dunning-Kruger effect*. In their original experiment, published in 1999, Cornell University psychologists Justin Kruger and David Dunning set out to test the hypothesis that incompetent individuals might tend to overestimate their ability and performance compared to objective measures.[34] Using Cornell undergraduates enrolled in psychology courses as their research subjects, Kruger and Dunning administered tests of logic, grammar, and humor and asked the students to rate their abilities in these domains relative to their classmates. Across all three test domains, there was a disparity between subjects' self-assessment of their ability versus their objective test performance among all but the top quartile of performers. That disparity increased as performance decreased, such that the lowest quartile displayed the disproportionately largest amount of overconfidence.

The Dunning-Kruger effect has been criticized as a mere statistical artifact and a misleading byproduct of how its authors graphed out their results in their original paper.[35] But, unlike other social psychology experiments that have crumbled under close scrutiny or attempts at replication, the Dunning-Kruger effect has been demonstrated time and again independently of how the original results were reported, among various types of people in both the psychology lab and in the real world, and across many different types of knowledge and task performance.[36] So, there's good evidence to support that the Dunning-Kruger effect is real, if sometimes misinterpreted or misreported by those who cite it. It does not, for example, tell us that only incompetent people are overconfident or that incompetent people believe they know more than experts. That was never what Kruger and Dunning's data showed: in their original study, subjective competence was rated as above average in all quartiles of objective performance. It's just that the objectively lowest performers had the most unwarranted confidence in their abilities, which is just what we'd expect if the Dunning-Kruger effect is at least partially a reflection of the better than average effect.

Another misunderstanding of the Dunning-Kruger effect is that it says more about incompetence than it does about some mythical category of "incompetent people" since objective competence and incompetence—that is, measurable knowledge about a given subject—varies within individuals according to topic or task. For example, an individual might be a world-renowned expert in quantum mechanics but completely incompetent in auto mechanics. In fact, we all have islands of relative competence and incompetence so that the Dunning-Kruger effect applies to all of us within the areas where we lack expertise. Dunning puts it this way: "Poor performers—and we are all poor performers at some things—fail to see the flaws in their thinking or the answers they lack"[37] and "not knowing the scope of [our] own ignorance is part of the human condition."[38] Put even more simply, we overestimate our abilities because we don't know what we don't know.

Dunning has always believed that his eponymous effect is best explained by a kind of "anosognosia of everyday life" or a blind spot reflecting a deficit in "metacognition," defined as the ability to evaluate our own cognitive processes such as knowledge, reasoning, and belief. His subsequent research has indeed clarified that this blind spot to incompetence is one that we're most likely to encounter when we look in the mirror—overconfidence is not, as a general rule, the result of underestimating the competence of others.[36] In addition, as subsequent research has shown, this lack of specific self-awareness persists despite performance feedback and can interfere with our ability to learn.[39] This isn't particularly surprising, since we might very well expect that people would be less likely to study or seek out new information about a given topic if they already thought they knew everything or even just more than the average person.

In 2018, Dunning and his co-investigator Carmen Sanchez from Cornell University also published evidence to support the adage that "a little bit of learning is a dangerous thing." In their experiments, they found that overconfidence was most likely to occur when people possess small amounts of knowledge about a subject as opposed to none at all.[40] Consequently, Dunning has suggested that while improving our powers of self-assessment might be a worthwhile goal, the best and perhaps only way to really fix unwarranted confidence is by "drinking deep from the Pierian spring"—that is, through attempts to more substantially improve competence so that our confidence is deserved.

Incompetence and the better than average effect aside, the most interesting thing about the Dunning-Kruger effect may be what it tells us about expertise. In their original experiment, Kruger and Dunning's results showed that, unlike everyone else, those in the highest quartile of test performance tended to modestly *under*estimate their performance compared to others. In contrast to those with less competence, this effect was attributable to both underestimating their own performance, suggesting a kind of "imposter syndrome"—the feeling we sometimes get that we're frauds who don't deserve the recognition we're given—as well as overestimating the performance of others.[32,36] In explaining this result, Dunning has referred to a quotation by American author William Feather (the same quotation has also been attributed to the French novelist Anatole France):

> An education isn't how much you have committed to memory, or even how much you know. It's being able to differentiate between what you do know and what you don't.[37]

This statement is, in turn, a contemporary version of the so-called *Socratic paradox* or the idea that true wisdom involves "knowing that one knows nothing."

Research on the Dunning-Kruger effect has consistently supported a smaller gap between expertise and confidence at the high end of competence compared to the low end, but the original finding that expertise is associated with underconfidence has not always survived replication.[36] In fact, experts can be as vulnerable to overconfidence as everyone else—not only outside of their area of expertise, as we would expect through the Dunning-Kruger effect, but within it as well. Based on their own research, Yale University psychologists Matthew Fisher and Frank Keil have found that experts are most vulnerable to overconfidence when they forget previously learned information that's necessary to maintain "formal" expertise based on extended study of a given topic.[41] This suggests that while objective competence may wane over time, in keeping with the dictum "use it or lose it," the confidence or "curse" of expertise, as Fisher and Keil put it, has a way of overstaying its welcome. It would seem that this "illusion of knowing" reflects a reluctance on the part of experts to admit their ignorance around topics they're supposed to know something about, especially in situations involving social pressure.[42]

If experts aren't immune to overconfidence, as Tversky and Kahneman also demonstrated within behavioral economics, or to cognitive biases in general, as other research has made clear,[11] we might therefore think of the kind of intellectual humility detected among experts in Kruger and Dunning's original research as representing another Goldilocksian ideal to which we should all aspire. As the Socratic paradox suggests, it may be that when experts really do know a subject well, they also understand the limits of the state of knowledge—not only their own, but in general—around that subject. When that happens, experts can be "confident, not cocky" and under certain circumstances, more willing to accept that limitation and admit, "I don't know." I'll have more to say about this in Chapter 9.

In the meantime, the Dunning-Kruger effect—along with behavioral economics, positive illusions, and the fallibility of memory—serves as a potent reminder that much of the confidence that we have about what we believe is likely to be unwarranted. While such overconfidence can make us feel better about ourselves, we would do better to acknowledge our tendency for ignorant self-deception and use that awareness to develop self-esteem that's earned. Indeed, the major benefit of looking in the mirror and coming to terms with what we don't know is that it can motivate not only experts but also all of us to learn.[42]

Still, achieving such humility is easier said than done. What would it really take to move ourselves closer to the Socratic notion of expertise? Remember Kahneman's statement on overconfidence—he fantasized about a having a magic wand to fix it. Dunning has a better reply:

> How do you get people to say, "I don't know"? I don't know.[38]

If only more of us were better at saying those three words instead of insisting that we're right all the time.

3
Confirmation Bias on Steroids

> Where self is, truth is not. Where truth is, self is not.
> —Buddha

Peripheral Brains

Thus far, I've been making the case that naïve realism—that is, overconfidence in our own subjective intuitions, experience, and worldviews—is a major cognitive pitfall that puts all of us at risk of holding on too tightly to false beliefs while stubbornly insisting that we're right. But, as I noted in Chapter 1, beliefs don't develop within a vacuum of subjective experience alone—much of what we believe is based on what we hear, read, or otherwise learn from others. Consequently, much of the rest of this book will be devoted to how we come to form beliefs through an interactive social process of alternately embracing or rejecting information that's out there in world. In this chapter, I focus on how the modern availability of information on the internet impacts that process.

Over the past several years, Matthew Fisher, the psychologist whose work on the overconfidence of experts I cited at the end of the previous chapter, has been conducting research on the potential for the availability of information online to boost our already hypertrophied sense of confidence about what we think we know. In 2015, he and his Yale University colleagues Mariel Goddu and Frank Keil demonstrated in a series of experiments that the act of searching the internet for "explanatory knowledge"—independently of finding it—inflates our confidence in what we think we know beyond what we actually know.[1] This seems to occur because we conflate the information that we recognize as accessible through our computers and cellphones—our "peripheral brains"—with knowledge that's stored in our actual brains inside our heads. It's as if our confidence in being able to retrieve information through what Fisher and his colleagues call a "transactive partnership with the internet" serves as a kind of proxy for real knowledge. Although they note that such a partnership could benefit us by increasing "cognitive self-esteem" in the same way that a positive illusion does, unwarranted confidence in what we know may come at a significant cost. As a result, mainstream news coverage of this finding has referred to it forebodingly as the "Google delusion."[2]

I distinctly remember witnessing the potential toll of peripheral brains on learning for the first time in my career as an educator back in the 1990s. In medicine, there's a long tradition of Socratic questioning to test the knowledge of students; one year, after the medical school decided that all students should be equipped with personal digital assistants (PDAs), I posed such a question during rounds. Instead of attempting to give an answer or to think it through, the student reached into his pocket, pulled out his PDA, and proceeded to retrieve the answer that way. The moment has always stuck with me—needless to say, I wasn't interested in testing the student's ability to locate the information within his peripheral brain; I was trying to determine if the information was in his real brain and, if not, to provide an incentive for him to put it in there.

I should say that while memorizing facts is sometimes useful and necessary, I think it's often overrated and I could barely stand it when I was a medical student, especially coming from an undergraduate education at MIT, where applying principles to engineer solutions was always encouraged over rote memorization. And, in recent years, the Socratic questioning of medical students in front of their peers, traditionally referred to as "pimping," has been criticized and relabeled as "toxic quizzing" that amounts to a kind of public shaming.[3] But if a significant benefit of peripheral brains is to offload the burden of memorization, the potential harm is that the confidence that it instills about what we think we know is at best unwarranted and at worst potentially harmful. For one thing, the instant access to information on the internet that we take for granted can just as easily vanish in an instant when our cellphone batteries die, when we're out of range of service, or when digital networks are hacked. For another, as I'll discuss in the next two chapters, there's no guarantee that online information is accurate, especially as text generated by artificial intelligence (AI) and digitally manipulated audio and images (e.g., "deepfakes") become increasingly common. And finally, we should all be concerned—as I was with my student—about how an overreliance on our peripheral brains can hamper learning.

As I mentioned in Chapter 2, Fisher's conclusion that experts may succumb to overconfidence due to forgetting previously acquired information provides validation for the "use it or lose it" dictum that applies to certain types of knowledge or skills. Although the tide has turned against "toxic quizzing" in medical education, the necessity of reinforcing knowledge through repeated testing is still well-recognized. For example, as a physician working in a hospital, I'm required to repeat cardiopulmonary resuscitation (CPR) training every three months. While such training is often viewed as an annoyance, I must admit that repeated training and performance testing does wonders to reduce any overconfidence in what I think I know while reinforcing the actual knowledge necessary to perform an urgent and potentially life-saving task.

In their more recent research, Fisher and his colleagues have shown that the awareness of information being accessible online not only results in overconfidence about what we think we know, but can also reduce our motivation to learn and the time we spend studying, as well as how we process information that we're retrieving online.[4] Whether consciously or unconsciously, it's as if we tell ourselves that it's not worth doing the work to put knowledge into our heads since we know we can retrieve it from our peripheral brains whenever we need to. That lack of incentive to learn is exactly what gave me pause when I witnessed my student try to recite an answer from his PDA, but that concern pales in comparison to what worries me now. As if boosting the kind of overconfidence in what we believe that I described in the previous chapter wasn't enough already, the added disincentive to learn new information and the potential for the "Google delusion" to interfere with information processing itself creates a perfect storm that further amplifies our vulnerability to embrace and hold tight to beliefs that aren't true.

Confirmation Bias, Confirmation Bias, Confirmation Bias!

In real estate, it's said that the most important guide to follow when buying a house and trying to understand home values is "location, location, location." If I were asked about the most important guide to understand the psychology of believing strongly in things that aren't true, I would similarly answer, "confirmation bias, confirmation bias, confirmation bias."

Confirmation bias refers to the tendency we all have to seek out, cite, or recall information that supports our preexisting beliefs and intuitions while ignoring or rejecting information that contradicts it. This tendency has been observed and remarked upon for centuries but was first modeled as a cognitive bias by Stanford University psychologists Charles Lord, Lee Ross, and Mark Lepper in 1979.[5] In a series of experiments, they presented research subjects with evidence to either support or refute the deterrent effect of capital punishment. Despite exposure to identical evidence, the research subjects reported a strengthening of their respective positions on the controversial topic due to accepting the presented evidence at face value when their position was supported and searching for alternative explanations and flaws in the data when it was refuted. This "biased assimilation" of evidence therefore amounts to alternately using or refuting objective evidence to support our preexisting intuitions and only counting as evidence that which supports what we already believe. Which of course is the exact opposite of what we should be doing: that is, dispassionately considering the available evidence to modify our beliefs over time rather than twisting it in the service of our immutable beliefs.

Like the Dunning-Kruger effect, confirmation bias has been well-replicated in other experiments and appears to be a universal bias to which we're all vulnerable, such that even scientists, who we might hope would be among the most objective and dispassionate collectors of data, aren't immune.[6] It's one of the most powerful determinants—if not the most powerful determinant—of how we consolidate our beliefs when sifting through the evidence as well as how we interact with others around disputed beliefs. When we disagree with other people about a given topic, confirmation bias predicts that we'll not only defend our beliefs by cherry-picking between different sets of data, we'll also insist on different interpretations of the same data. This of course means that disagreements often don't get us anywhere because we can't agree on something as fundamental as what counts as evidence. And, even worse, we are for the most part unwitting victims of confirmation bias—we're unable to see our own vulnerability to it but eager to attribute it to others who disagree with us, especially now that it has become a household word. This tendency to attribute cognitive biases to others while failing to acknowledge them in ourselves represents a form of naïve realism that's known as the *bias blind spot*.[7]

In the internet age, confirmation bias now means something rather different than it did when Lord, Ross, and Lepper wrote about it in the late 1970s. When I was growing up in that era myself, it was hard to fathom my parents' own childhood without television, huddled around the radio—for some reason, I always imagined them listening in the dark—when their favorite programs would come on air. But now I'm at the age when it's my own son, and even my students in their twenties, who struggle to appreciate what the primitive state of technology was like when I was a kid. In elementary school, I was given "book reports" and other writing assignments that required biking down to the local library to find a single volume on a given subject—often with the help of a librarian—or consulting hard-bound encyclopedias as consolidated sources of authoritative knowledge. I wrote out my homework by hand and later used a typewriter and products like White Out and Liquid Paper to paint over mistakes. If I ever wanted to move a paragraph of text around, I had to start over from scratch. When my family watched TVs with large cathode ray consoles, "rabbit ear" antennae, and no remote control, we had three major networks to choose from and looked to a select group of anchors like Walter Cronkite, Dan Rather, Ted Koppel, Peter Jennings, and Tom Brokaw for our trusted sources of daily news. In college, I called my parents on telephones plugged into a wall no more than every few months because long-distance calls were prohibitively expensive. I mailed handwritten love letters to my long-distance girlfriend and eagerly awaited replies over the course of weeks at a time. Even during medical school and residency training, a mere 30 years ago, I spent considerable time and money in the library stacks using a Xerox machine to copy journal articles from bound volumes when they weren't checked out or missing.

Suffice it to say that the world is different now, with computers, word processors, cellphones, and the internet having radically transformed how we access, obtain, and share information. We enjoy nearly instant access to a much wider scope of information than I could have ever retrieved from the local library as a child, along with countless news options from both cable TV and online sources, all of which can be shared with others in an instant. The sheer volume of information that's now readily available at our fingertips makes confirmation bias much more likely because it's easier than ever to find information that supplies the "evidence" to support almost any belief while "swiping past" information that contradicts it.

Delusional Thinking Meets the Internet

Some years ago, I was treating Frank Dunbar, a thirty-something-year-old man with schizophrenia who was convinced that government satellites were deliberately bombarding his body with "energy beams" that were causing him physical pain and making his limbs twitch. Over the course of several weeks in the hospital, I attempted to persuade him to take antipsychotic medication, but, as is often the case with those suffering from delusions, he didn't think it would be helpful because he insisted that he wasn't mentally ill. With time, patience, and persistence of his mental suffering, his refusal gradually abated, and, one day, he finally agreed to give the medication a try. Unfortunately, just as he'd done so, a new patient was admitted to the unit and spoke up in a group therapy session to say that she, too, was being attacked by satellites. On hearing that, Mr. Dunbar's eyes lit up and he said, "See? I *knew* it was true!" and he promptly refused to take medications again. We were back to square one.

In Chapter 1, I wrote about how delusions are generally regarded as unshared and even unshareable beliefs, but since delusions often follow stereotyped themes, their overarching premise—in Mr. Dunbar's case, the idea that the government was zapping people with energy beams—can sometimes overlap between patients. Still, that kind of communality was less likely to occur outside of a psychiatric ward before the internet existed. A hundred years ago, you might search an entire town and still not find anyone willing to subscribe to your unconventional belief—you'd more likely face ridicule or be referred to the asylum. But now the internet has made it easy to find others who share even the most unconventional, idiosyncratic, and fringe beliefs in a way that Karl Jaspers or the authors of the *Diagnostic and Statistical Manual of Mental Disorders* (DSM) who defined delusions could have never anticipated. In today's virtual world, where you can reach like-minded strangers on the other side of the planet at the click of a button, the premise of a "fringe belief" barely makes sense anymore. The "fringe" isn't lurking somewhere in the hinterlands; it's right there in the palms of our hands. As a result, it's not unusual these days to find instances of delusional beliefs that are fueled by online support.

Take, for example, delusions of infestation or *delusional parasitosis*—the false belief that one is infected by bugs, parasites, or some other organism. Such delusions have been well-recognized in medicine for centuries, with sufferers often presenting to primary care providers or dermatologists for help. About 20 years ago, a mother whose son had been diagnosed with delusional parasitosis proposed that an actual infestation with mysterious "fibers" that she dubbed "Morgellons disease" represented a better explanation. She formed the Morgellons Research Foundation out of her home to promote awareness and generate funding for research, and, with the help of the internet, "Morgellons" became increasingly well-known and covered in the media. As a result of rising public concern about this supposedly novel medical threat, the Centers for Disease Control and Prevention (CDC) was persuaded to study 115 suspected cases in Northern California, where there had been an alleged outbreak of the condition. In 2012, the CDC published its study results, which concluded that there was no evidence of infection and that the only fibers detected were likely cotton strands from clothing.[8] Today, Morgellons is widely understood within the medical community as a form of delusional parasitosis "facilitated by web-based dissemination" and an "internet meme,"[9] but it lives on as a real infectious disease in the minds of purported sufferers as well as in a select few researchers who have paired Morgellons with their previously established interest in chronic Lyme disease, another medically disputed syndrome.[10]

In much the same way, paranoid delusions have been animated in recent years by online support for a phenomenon called "gang-stalking" based on claims that self-described "targeted individuals" are being routinely followed, surveilled, harassed, and otherwise victimized by malevolent forces wielding high-tech weapons of "mind control" using the likes of "extremely low-frequency" radiation or "voice to skull" technology.[11] In 2006, Cardiff University clinical psychologist Vaughn Bell and his colleagues published an analysis of 10 online accounts of "mind control experiences" consistent with gang-stalking. When assessed by three independent psychiatrists, all of the accounts were rated as delusional.[12] A decade later, Curtin University psychologist Lorraine Sheridan and David James from the National Stalking Clinic in London conducted an analysis of 128 responses to a survey about stalking that likewise concluded that all of the cases involving alleged gang-stalking by coordinated groups of people reflected paranoid delusions, in contrast to the 96% of claims about stalking by single individuals that were deemed to be real.[13] In both of these studies, gang-stalking claims were attributed to paranoid delusions because they defied credibility due to the sheer amount of resources or level of coordinated organization that would be necessary to carry out what was claimed. In short, the beliefs were unshareable, at least almost anywhere other than on the internet—where anything goes.

To be clear, stalking, bullying, online harassment, and "mobbing" as well as zoonotic infestations by bedbugs, fleas, mites, and other parasites are real-life problems that cause people considerable distress. But delusions of persecution and parasitosis also exist, and, due to the proliferation of personal accounts and unsubstantiated claims about the likes of Morgellons and gang-stalking online, delusional thinking now risks being conflated with real-life events on a scale that wasn't possible before the internet. On the bright side, those who self-identify as having Morgellons or as targeted individuals can find solace within online spaces where they can share "war stories" and survival strategies with like-minded individuals.[14] But the dark side of the internet is that it has also created a forum for false information and subjective testimonials to masquerade as objective evidence, even when good evidence to the contrary exists, steering people away from real answers and psychiatric treatment that might prove helpful.

* * *

But wait a minute, you might ask having read Chapter 1, are delusional beliefs sharable or aren't they? If all manner of beliefs can now be shared online, how can we ever distinguish delusions from more normative beliefs in the internet age? To answer such questions, it's necessary to examine beliefs on a case-by-case basis, with careful attention to the underlying evidence used to defend them. Let's consider four hypothetical case scenarios of strongly held beliefs.

In the first scenario, recall Mr. Roberts, whom I described at the start of this book. He believed that an "entity" living inside his body could control his actions because he often felt his body move without his conscious control or any desire to act. His false belief was about himself, was based on subjective experience without any objective evidence to support it, and was unshared by others, so it would fall comfortably into the category of a delusion.

In the second case, let's say that Mr. Roberts had a son named Deron. Ever since he was a young boy, Deron listened to his father recount stories about the "entity" and heeded his warnings that the same fate might befall him one day. Over the course of his childhood, Deron came to accept what Mr. Roberts taught him as gospel. Then, one day, after talking about the risk of entities inhabiting people's bodies in his fourth-grade health class, Deron is referred to the school psychologist for an evaluation. Since his belief is about the state of the world rather than himself and is largely based on the delusional testimony of his father, he would be appropriately diagnosed with *folie-à-deux* or what the DSM used to call "shared psychotic disorder." As I mentioned in Chapter 1, such examples of shared delusions are relatively rare but have been described at least as far back as the 1800s and typically occur within close relationships such as families, marriages, or "cult" environments where impressionable individuals adopt the delusional thinking of someone with a psychotic disorder like schizophrenia. In such cases, the treatment of the impressionable individual typically involves separating and "deprogramming" them from the influence of the "primary case" (i.e., the person who convinced them of the delusional belief).

In the third scenario, we return to Frank Dunbar, who believes that there are government satellites zapping energy beams into his body. He believes this because he experiences his neuropathic pain as a burning sensation and he mistrusts the government, but also because he heard another patient on the psychiatric unit describing similar experiences. After discharge from the hospital, Frank later stumbles upon YouTube videos about gang-stalking so that he finally feels like he's found an explanation for his experiences that validates what he's been telling psychiatrists all along—that he's not crazy. His belief is about himself and is based on subjective experience, but the overarching persecutory theme is shared by others online. This would be an example of a delusion amplified by similar testimonials found on the internet.

Finally, let's consider a fourth case in which Cecily Perkins, a small-town store manager, has also read a stack of books and watched hours of YouTube videos about gang-stalking. She doesn't believe that she's a victim herself but is worried that she could be if it's true that law enforcement agencies, government institutions, and private corporations are routinely using "voice-to-skull" technology to harass innocent victims for whatever reason. She posts on social media demanding a halt to such violations of civil liberties and writes her congressman to take up the cause. Since Cecily's belief isn't about herself and because the evidence to support it is based on what she's read and watched online rather than subjective experience, she shouldn't be considered delusional. Her beliefs also shouldn't be considered an example of shared psychotic disorder since, unlike Deron, she learned about gang-stalking from published books and videos, not from a lone individual with delusions. As beliefs that bear a superficial resemblance to delusions but fall short on several fronts, they fit comfortably under the broader category of delusion-like beliefs that shouldn't be conflated with evidence of mental illness.

It ought to be clear by now that making a clinical decision about whether a belief is delusional or not and deciding whether or how to treat it depends on a number of factors, including the self-referentiality of the belief, the kind and quality of evidence to support it, and the quantifiable cognitive dimensions of belief such as conviction and preoccupation that I discussed in Chapter 1. But, as always, there's a gray area between clear-cut delusions and delusion-like beliefs, and, in the internet era, the fourth scenario in which people without mental illness, like Cecily Perkins, come to develop fixed and false but shared beliefs about the world seems to be increasingly common. This can be a source of significant confusion not only in clinical work, but also in legal arenas when people commit crimes based on such beliefs.[15]

Note, however, that while it's not uncommon to hear people apply the term "delusion" loosely to describe dubious beliefs unsupported by evidence that are shared by groups of people and even entire cultures—like the age-old claim that religion represents mass delusion—that sort of shared belief is precisely what the DSM excludes from its definition, as I explained in the previous chapter. Additionally, while I explained in the Preface and Chapter 1 that delusions can be conceptualized as lying at one end of a continuum or spectrum of false belief, it's still useful to separate them from the less pathological end of that spectrum, just as doctors distinguish high blood pressure from normal blood pressure. And while it could be argued that using the term "delusion" loosely outside of psychiatry is justifiable in the same way that the term "cancer" is used as a metaphor in nonmedical discourse, few of us would ever confuse a metaphorical cancer with actual cancer. Since that's not the case when the term "delusion" is applied carelessly, it's crucial that we steer clear of using terms like "delusion" or "psychosis" to describe widespread false belief, especially when they're applied as a pejorative to those whose beliefs conflict with our own.

Confirmation Bias on Steroids

Some years ago, I came across the following internet meme:

> Let's face it, if you were born in Israel, you'd probably be Jewish. If you were born in Saudia Arabia, you'd probably be a Muslim. If you were born in India, you'd probably be a Hindu. But because you were born in North America you're likely a Christian. Your default faith is not inspired by some divine, constant truth. It's simple geography, and adapting to the faith of your family.[16]

Though the quotation seriously short-changes religious diversity that's long been present in the world, it nonetheless resurrects a historical truism about how many if not most of our cherished and identity-defining beliefs and values have been inherited as geographical "accidents of birth" from the families and subcultures in which we were raised or that we subsequently inhabited.

Because the internet has since transformed the geography of belief, freeing it from its previous cultural and familial confines into the vastness of virtual space online, we might expect that we've become more freethinking today. But no such luck: while we might be more likely to disagree with our family members, we've mostly remained unable to escape our tendency to become sequestered into ideological tribes and may have merely exchanged our traditional geographical and familial tribal boundaries for new digital "echo chambers."

Over the past decade, increasing light has been shed on how this occurs in the internet era today. Back in 2011, Eli Pariser coined the term "filter bubble" to describe how the digital architectures of search engines and social media sites have been designed to feed us what AI algorithms "think" we want to see based on our previous clicks, comments, and searches online.[17] Simply put, because clicks generate revenue within online spaces, internet sites are increasingly designed to get us to click more. And so, whether we're looking for news and videos on Google and YouTube and even when we're just scrolling through our daily feeds on Facebook and Twitter, each of us is fed a personalized view of the world that's been crafted to grab our attention. From an informational standpoint, this steady diet of tailored content amplifies our already existing cognitive bias to click on information that confirms our worldview while ignoring that which contradicts it. The result? Confirmation bias on steroids and the potential for two people to come away from the internet with two completely different accounts of reality. Indeed, it has been suggested that, like peeking at someone's secret diary, examining someone's viewing history on YouTube might give us the most unfiltered and reliable idea of what someone is likely to believe.[18] By looking at someone's queue of recommended YouTube videos, we might even be able to predict what some people will come to believe in the future.

The best depiction of confirmation bias that I've ever come across is a three-panel comic strip by cartoonist Kris Straub.[19] In the first panel, a man is sitting in front of his computer and thinks to himself, "I've heard the rhetoric from both sides ... time to do my *own* research on the real truth." In the second panel, we see the man's computer screen where he has typed in "hotly debated topic" into a Google-like search engine. His cursor points to the first of 80,000 hits entitled, "literally the first link that agrees with what you already believe." Then, in the third panel, we see him peering over his glasses to eyeball the screen while he says, "jackpot." This is a perfect illustration of how confirmation bias holds new meaning in the internet age and why Dan Brown, in his bestselling novel *The Lost Symbol*, cautioned that "Google is not a synonym for 'research.'"[20] But due to the filter bubbles that we encounter online, someone like the man depicted in Staub's cartoon doesn't only click on the first hit that confirms his belief and stop there—he's fed pages and pages of content that allow him to come away with the impression that the evidence to support his beliefs is overwhelming. That's confirmation bias on steroids.

Another real-life story of online confirmation bias gone awry appeared in a 2016 *Washington Post* article by Stephanie McCrummen entitled, "'Finally. Someone who thinks like me.'"[21] In it, McCrummen paints a revealing portrait of Melanie Austin, a retired railroad worker living in the small town of West Brownsville, Pennsylvania, who found "evidence" online to support her suspicions about a wide range of fringe political beliefs—that President Obama was gay, a Muslim, and the co-founder of ISIS along with Hillary Clinton; that Michelle Obama was actually a man named Michael; and that Supreme Court Justice Antonin Scalia was murdered by a prostitute. These beliefs were both sparked and fueled by watching countless YouTube videos, reading right-wing political sites that came up on Google when she searched for topics like "Scalia murdered by prostitute," and by having amassed thousands of Facebook friends and Twitter followers whom she was "meeting every day across America" and that "felt the same way as she did." After posting a self-described "rant" about how President Obama should be hanged and the White House burnt to the ground, she was involuntarily hospitalized for a psychiatric evaluation and discharged after several weeks with no diagnosis beyond "homicidal ideation." Looking back on the ordeal, she concluded that it was unjustified and outrageous, noting that, "If it's time to lock me up, it's time to lock up the world." Just how could someone like Austin arrive at such a conclusion? Because she wasn't only searching and finding evidence to support her pet theories: she was having them served up to her on silver platter courtesy of online filter bubbles and echo chambers, giving her a distorted and narrow view of the world and egging her on. Two years later, Austin would go on to be charged with a string of criminal offenses including terrorist threats, reckless driving, disorderly conduct, and criminal mischief.[22]

Pariser and others like him expressed concern early on that being relegated into online echo chambers might put us all at risk of reduced exposure to different points of view so that we might end up digging in our heels about our beliefs more than ever. More than a decade later, it's now clear that personal beliefs can indeed be reinforced through confirmation bias on sites like Facebook and YouTube and that this can in turn result in a greater ideological polarization.[23] But contrary to what was initially feared about filter bubbles and echo chambers, ideological polarization doesn't necessarily occur due to lack of exposure to opposing views. In fact, we do routinely encounter the viewpoints of our ideological opposites both online and through other informational sources,[24] but, in keeping with the mechanics of confirmation bias, we tend to simply dismiss or reject them when we see them.

It therefore appears that online echo chambers aren't so much defined by users wearing passive digital blinders but by the act of discounting informational sources that contradict our worldviews. Utah Valley University philosophy professor C. Thi Nguyen therefore urges us to distinguish between the kind of "epistemic bubbles" that Pariser warned about, where we aren't exposed to dissenting voices in online spaces in the first place, and "echo chambers" proper, where dissenting voices are actively discredited and undermined due to lack of trust in informational sources coming from the other side of the ideological fence.[25] He further reframes the phenomenon as not so much succumbing to irrationality, faulty reasoning, or even cognitive biases so much as just following the instructions of engagement set up by the echo chambers that we inhabit online. After all, why should we be expected to heed informational sources that we've been taught not to trust? This argument suggests that people today often find themselves hapless ideological victims of digital destiny, just as we did when we formed our beliefs according to geography and familial boundaries in days of yore.

Flame Wars

Even if there's something of a rational innocence to our dismissal of counterevidence within online echo chambers, as Nguyen suggests, this seems like an undeserved defense of collective naïve realism that shortchanges the role that confirmation bias plays in putting people into echo chambers in the first place and driving them deeper into conviction around their beliefs once there. I find it particularly hard to give ourselves—and the internet, where we often now spend more time interacting with people than we do face to face—a pass for how our behavior within online echo chambers puts us at odds with, and at the throats of, our friends, family, and strangers alike. When Lord, Ross, and Lepper first studied confirmation bias in 1979, before the internet existed, they concluded that confirmation bias resulted in "attitude polarization"—a widening of the gap between two sides of a controversial topic.[5] With confirmation bias on steroids, polarization isn't just two sides set apart from each other and vulnerable to self-deception—it's as if the two sides are magnets that the internet has flipped around so that their repulsive forces make it impossible for them to ever meet in the middle.

When I was in high school, one of my classes required that students participate in regularly scheduled group debates. We were assigned to argue a specific side of a topic, with our grades hinging on the teacher's judgment of the strength of the cases we presented. Needless to say, being a disaffected teenager, I found the exercise frustratingly pointless because the grading meant that one side would never concede that their opponent's argument made more sense or consider accepting a more neutral, balanced view. On the contrary, earning an "A" seemed to be contingent on making a strong argument, sticking to it, and never backing down despite listening to sensible counterevidence. What, I thought, was the point of such a fool's game? Over the past few years, my interactions on social media sites like Twitter have often left me asking the very same question.

When people encounter counterevidence and dissenting opinions online, they're often not merely ignoring, dismissing, or "swiping past" according to the stubbornness of confirmation bias—they're engaging with their ideological opposites out of a desire to fight. A growing body of research has confirmed this, at least within the narrow category of online political discourse over the past several years. For example, based on their analysis of Twitter interactions from 2017 around the topic of then-president of the United States Donald Trump, Northeastern University researchers Sarah Shugars and Nicholas Beauchamp found that back-and-forth engagement with those holding opposing views was common but often amounted to "acrimonious argumentation" that was most likely to occur when Tweets consisted of negative sentiments and "unpleasant words."[26] Other studies have likewise found that both engagement in political debates and the sharing of political information on social media are best predicted by anger.[27] Researchers at Yale analyzed 12.7 million politically related Tweets across 7,331 Twitter users and found evidence that expressions of "moral outrage"—defined as reactions of anger, disgust, and contempt and statements of blame or a desire to punish in response to perceived violations of personal morals—are rewarded through both "reinforcement learning" and "norm learning."[28] Users were more likely to get "likes" and "shares" when they Tweeted expressions of outrage—especially within networks in which such expressions were the norm—and, when so rewarded, they were more likely to Tweet out more of the same. A similar study comparing 6.5 million comments on news articles posted on Facebook pages to responses to a national survey found that "toxic" comments were 77% more common on Facebook, where they attracted more "likes" than civil comments and where exposure to them increased the toxicity of subsequent commentary.[29] The researchers described their findings as evidence of an "incivility contagion" created by the "distorting prism of social media" that amplifies people's hostilities toward one another. Finally, a study of social media posts on both Facebook and Twitter concluded that the most potent predictor of sharing or retweeting political content from media outlets and politicians was negative emotional commentary specifically directed at our ideological opposites, leading the researchers to conclude that "out-group animosity drives engagement on social media."[30]

What does all this research mean? It means that, on social media, anger toward other people and their ideas generates clicks. And since clicks generate revenue, such findings have not been lost on the architects of social media platforms.[31] In a yet to be published study by researchers from Cornell Tech and University of California Berkeley, the use of curated content on Twitter that generates tweets selected "for you" as opposed to showing tweets from users that you're "following" increased exposure to tweets from out-groups as well as expressions of out-group animosity while increasing negative feelings toward our ideological opposites in the process.[32] Indeed, I experienced this firsthand on Twitter a few years ago when the "for you" feed activated itself on my account one day, showing me a stream of tweets from people I would just as soon not hear from and filling me with a palpable sense of what I can only charitably describe as annoyance. After I figured out what was going on, I promptly switched back to the "following" feed where I make it a point not to block people who argue with me and to follow at least some accounts with whom I disagree. For me, this happy medium ensures that I don't find myself too confined within an echo chamber while allowing my usual level of annoyance on Twitter to be merely occasional as opposed to constant.

Such is the dark side of getting out of our epistemic bubbles. As Nguyen suggests, navigating social media within echo chambers doesn't mean that we don't encounter or interact with our ideological opposites. We do—in fact, it's unavoidable. But the result is more often annoyance or antipathy than understanding or any softening of our ideological convictions. And all the while, the social media companies are stoking our ire and exploiting it for profit.

* * *

I'll tackle the thorny topic of political polarization further in Chapter 8, but, for now, the more general implication of this research is that the algorithms and dynamics of social media platforms are constantly egging us on, incentivizing us to fight with each other when we're online. So there's good reason to conclude that angry disputes are more likely to occur on the internet—where we speak of dominating our ideological opponents by "owing" or "pwning" them—than they are when we interact with people face to face. But we can't put all the blame on social media for such angry disputation. There are several other psychological reasons why we're more likely to find ourselves taking the bait and fighting "flame wars" with strangers when we're online.

For one thing, when we're presenting our digital personas to the world, we want to look good—to show everyone how happy, beautiful, and smart we are. And when we're angry and looking to pick a fight, the last thing we want to do is concede that we're wrong. So, just like in my high school debate class, that means never backing down from an argument and holding on ever more tightly to our beliefs. Keeping Matthew Fisher's research on peripheral brains in mind, arguing on the internet allows us to pose as someone with expertise, armed with the ability to search for information online as we're interacting. It's as if we're contestants on Jeopardy! who are looking up the answers on our cellphones in real time.

Armed with access to our peripheral brains, there's also no better place than online to "stick it to the man." The internet represents the ultimate egalitarian forum for sharing ideas, a place where everyone and anyone can offer an opinion just as easily as a voice of authority or power. Back in 2004, Rider University psychologist John Suler popularized the term "online disinhibition effect" to describe how such online egalitarianism might encourage people to come out of their shells to participate in public discourse.[33] As Suler and others have noted, this effect has the potential for social benefit—for example, by allowing people to share their inner emotions and giving a voice to those who might otherwise be too shy or feel too disempowered to speak up. But online communication can also facilitate "toxic disinhibition," where angry argumentation, personal attacks, cyberbullying, hate speech, and even threats flourish.

Another reason we're more likely to find ourselves fighting "flame wars" online than we would in person is due to a kind of "road rage effect." When we're driving in our cars alone and someone cuts us off in traffic, many of us find ourselves shouting expletives at the top of our lungs in a way that we'd never dream of doing face to face with the other driver. We do this because it gives us a chance to vent our anger in private while avoiding the kind of direct confrontation that might degenerate into a physical altercation or worse. In much the same way, Suler speculated that the disinhibiting effects of online communication result from the threefold ability to be anonymous (hiding our identities), invisible (not seeing those we're communicating with face to face), and asynchronous (not interacting in real time). Of these, anonymity seems to be the sharpest two-edged sword. On the one hand, research published in 2014 by Old Dominion University professor Russell Haines and his colleagues confirmed that anonymity can increase participation in online discourse. But, on the other hand, it does so across the board, without any "equalizing effect" that might disproportionately benefit those too shy to speak up.[34] Instead, Haines and his colleagues contended that anonymity "removes the accountability cues and frees members to express unpopular or socially undesirable arguments," "freeing reticent opinions" as opposed to reticent people. University of Houston professor Arthur Santana found further evidence of this when comparing online comments between news sites that allowed either anonymous or non-anonymous reader commentary.[35] Anonymous commenters were significantly more likely than non-anonymous commenters to make an "uncivil" comment consisting of a personal attack, vulgarity, ethnic slur, racist remark, or threat. A slight majority (53%) of anonymous comments were uncivil, whereas only about 30% of non-anonymous comments were uncivil.

It seems, then, that just like when we're alone in our cars, the anonymity of online communication gives us license to turn off our filters and speak our minds, sharing opinions that we'd more likely keep private—appropriately so—in face-to-face interactions. Except that we're not alone when we're online, so that when we're arguing with others on the internet, the result of incivility is to predictably worsen attitude or ideological polarization—disagreeing with others and moving us farther away from common middle ground.

It could be argued that this is more a problem of a few bad apples in the form of internet "trolls" and within particularly toxic online forums where they flourish like Reddit, 4chan, and 8kun. After all, not everyone on social media sites like Facebook, Twitter, and Instagram or the internet more broadly falls into the trap of online road rage. But information science researchers at the University of Wisconsin-Madison led by Ashley Anderson have found evidence for a "nasty effect" of online incivility even among "lurkers" or passive observers of online debates.[36] When research subjects reading a neutral and balanced news blog about the safety of nanotechnology were exposed to uncivil comments about the topic and its author, those skeptical about nanotechnology became more skeptical, whereas those supportive about nanotechnology became more so. No such attitude polarization occurred when the observed comments were civil. These findings suggest that we don't only dig in our heels about our beliefs when we're directly involved or attacked in an online argument as we might expect: we also do so when we're passive observers to the fray and even when perusing fairly neutral and unbiased informational sources. To return to my earlier metaphor, online incivility can flip the magnets so that we're pushed apart from our ideological opposites whether we're actively participating in a debate out of anger or just watching from the sidelines.

* * *

Can't we all just get along online? It seems like the answer might be "no" and that even when it's a smaller minority of trolls who are disrupting online spaces, the net result is to amplify attitude polarization for all of us just the same. If all we see is fighting, we come away believing that everything is endlessly debatable, that truth doesn't exist, and that we may as well just believe whatever we want to believe. Due to concern about such effects, a growing number of online news sources including Reuters, ESPN, *USA Today*, *Huffington Post*, *The Chicago Sun Times*, Yahoo!, *The Atlantic*, *The New York Times*, and *Psychology Today* (where I write a blog) have opted to eliminate reader commentary altogether in recent years, providing a timely illustration of the meme, "this is why we can't have nice things."

To be clear, the internet is a "nice thing," and I'm not suggesting that we all become neo-Luddites who call for its destruction or even that we should unplug from our peripheral brains. It has given us a lot to be thankful for—rapid access to breaking news is a major upgrade from the world in which I grew up, and social media has great potential to put us in touch with loved ones and strangers alike and to bring people together. I couldn't have written this book without the many sources of information that I was able to find and readily access online. But our gratitude for the internet should acknowledge the potential for this gift to be a destructive social force. As University of Virginia media studies professor Siva Vaidhyanathan warns, social media can often amount to "antisocial media."[37]

Within the click-based economy of the internet, all too often profits are maximized by generating content that exploits our cognitive vulnerabilities so that we're sucked into a new virtual world with the unintended consequence of amplifying the already unwarranted amount of overconfidence we have in our personal beliefs, many of which aren't true. Between confirmation bias on steroids created by filter bubbles and echo chambers and online incivility created by anonymity, it's now easier than ever to not only disagree, but hate our neighbors based on those false beliefs. Surveys indicate that we're aware of this and don't want it to be this way,[38] but, so far, that hasn't been enough to change things. In the next two chapters, I'll take a closer look at why that is.

4
The Flea Market of Opinion

> Falsehood flies, and the Truth comes limping after it; so that when Men come to be undeceiv'd, it is too late; the Jest is over, and the Tale has had its Effect.
>
> —Jonathan Swift

When Misinformation Kills

On a spring day in 1997, local sheriffs responded to an anonymous tip about a "mass suicide" at a seven-bedroom mansion in the affluent Southern California neighborhood of Rancho Santa Fe. After breaching the doorway, deputies were greeted with a pungent odor and found 39 men and women lying peacefully—and quite dead—in their bunkbeds, dressed in uniforms consisting of black shirts, sweatpants, retro Nike sneakers, and armbands along with purple cloths draped over their heads and torsos. The armbands bore patches that identified them as members of the "Heaven's Gate Away Team."

Heaven's Gate was a "new religious movement"—some would argue that the pejorative term "cult" was more appropriate[1]—founded in the 1970s by Marshall Applewhite and Bonnie Nettles, whose doctrine was based on the belief that mankind was spawned by extraterrestrials visiting the Earth millennia ago. Adherents of the movement were taught that they should transcend the physical body, which they called "the vehicle" and a mere "container," and aspire to attain a kind of spiritual ascension that they referred to as the "Next Level" or "The Evolutionary Level Above Human."[2] It's been said that at one time the membership of Heaven's Gate numbered up to a thousand people and that the group still attracts aspiring followers to this day.[3] Back in 1997, however, an evolving rumor about a spaceship hidden in the tail of the Hale-Bopp comet—which was due to make its closest pass by Earth in some 4,000 years—narrowed the core membership to 39 dedicated true believers who were eager to make the ultimate wager on the belief that by ending their corporeal existence here on Earth, they would finally reach the Next Level as their souls were transported onto the spaceship.

Recalling the discussion about delusions in previous chapters, the fixed, false beliefs shared by the members of Heaven's Gate involved a self-referential aspect that served as the motivational rationale to end their lives and deny that doing so would be an act of suicide. As such, the group's beliefs might be attributable to an example of "shared psychotic disorder" or "folie-à-39" that I mentioned in Chapter 3, particularly if it could be determined that Applewhite, the belief's progenitor who billed himself as a reincarnated extraterrestrial who had taken human form and was said to have been hospitalized psychiatrically back in the 1970s,[4] was delusional himself. But as I discussed in Chapter 1, religious beliefs (i.e., religious doctrines shared by other members of a person's culture or subculture) are typically excluded from the psychiatric definition of delusion, although the *Diagnostic and Statistical Manual of Mental Disorders* (DSM) leaves the inevitable question of how many people it takes to form a subculture unanswered. As idiosyncratic as the extraterrestrial doctrine of Heaven's Gate might sound, it has significant overlap with that of the much larger and well-known Church of Scientology. And, from an atheist's perspective, the belief that death might be a gateway to board an alien spacecraft isn't that much more of a stretch than the Christian belief that when we die, we might end up in Heaven seated at the right hand of God or eternally toiling in the fiery pits of Hell. It would therefore be problematic to label the shared beliefs of Heaven's Gate as shared delusions without similarly implicating a wide range of more mainstream religious beliefs along with them.

Even if the members of Heaven's Gate weren't delusional, the temptation to chalk up their fate to the deranged behavior of a millennialism cult might still be hard to resist. Indeed, when asked for comment at a news conference at the time, CNN founder Ted Turner cavalierly dismissed their suicides as "a good way to get rid of a few nuts."[5] But it becomes even more difficult to sweep the phenomenon of Heaven's Gate under the rug of mental illness when we consider that the evidence to support the belief that a flight to the Next Level was awaiting inside the tail of the Hale-Bopp comet wasn't only a shared belief that originated from the mind of Applewhite. Taking a closer look at its origin story, the claim that the UFO was hidden there had been popularized over the previous year on the late night talk radio show *Coast to Coast AM*, hosted by Art Bell, with photographs of the purported spaceship appearing as an amorphous white spot posted on Bell's webpage and elsewhere on the internet. So, unlike delusions based solely on subjective experience, the Heaven's Gate UFO beliefs were supported by objective evidence—except that within a short time after it appeared online, astronomers debunked the rumor and revealed the photographs to have been doctored as part of a hoax.[6] Although the members of Heaven's Gate were apparently privy to this debunking, there was no turning back. It was almost as if a loaded gun had been handed to the Heaven's Gate members who were intent on ending their lives and had finally found the validation they needed to go through with it. In the aftermath of the suicides, it was suggested that both fraudulent evidence and lack of journalistic integrity contributed to the Heaven's Gate members' deaths. Although Bell responded, "I'm not going to stop presenting my material because there are unstable people,"[7] the role that misinformation played in what ended up being lethal beliefs—whether it was coming from Marshall Applegate or Art Bell—shouldn't be minimized or ignored.

A study published at the end of 2023 revealed something that was both surprising and disturbing about confirmation bias on steroids and its effects on the oft-heard recommendation to "do your own research" in the search for truth that helps us to understand why the members of Heaven's Gate might have been so easily egged on by misinformation. Across multiple experiments, when those exposed to false or misleading news articles were instructed to conduct online searches to evaluate them, doing so *increased* rather than decreased the probability of believing the false articles to be true.[8] This appears to occur because when we search online about misinformation, we're likely to be exposed to more misinformation on that same topic within what the authors refer to as "data voids" and "propaganda feedback loops." Such a conclusion highlights that our efforts to decide what we believe or don't believe involves not just our brains, but also the interaction between our brains and the informational landscapes that we navigate in search of answers. In this chapter, I'll explore how the ubiquity of misinformation in the world today makes it all too easy to adopt and defend false beliefs by claiming that we have the evidence to back them up.

* * *

In recent years, there have been many other timely and more terrestrial examples of misinformation leading to violent behavior and tragic ends than Heaven's Gate. Two stand out as especially poignant object lessons of the modern era. The first is "Pizzagate." In 2016, when Edgar Maddison Welch was apprehended by police for firing his semi-automatic rifle inside the Comet Ping Pong pizzeria to "self-investigate" whether it was the headquarters of a then–US presidential candidate Hilary Clinton–led child pornography ring—a rumor he'd heard on conservative radio and the likes of *InfoWars*—he had to concede that "the intel on this wasn't 100 percent."[9] Nevertheless, the faulty "intel" that was the Pizzagate conspiracy theory left an enduring legacy beyond Welch, who served time in prison and was released four years later. A direct line can be drawn between Pizzagate and the subsequent rise of the QAnon conspiracy theory movement that culminated with the violence at the US Capitol in January 2021 (I'll return to the topic of QAnon and other conspiracy theories in Chapter 6).

The second example of belief in misinformation gone awry had few if any comparable ripple effects but is tragic in its own right. In 2020, celebrity stuntman and Guinness World Record holder "Mad Mike" Hughes staged an attempt to launch himself into the troposphere with a homemade steam-powered rocket so that he could verify his belief in a flat Earth. Although such belief regarding the shape of our planet once represented popular dogma grounded in naïve realism, it was largely abandoned as far back as the 1300s, such that most of us now take its falsehood for granted. And yet, flat Earth beliefs have reemerged in recent years based on evidentiary claims presented online in YouTube videos and popularized by celebrities like NBA basketball player Kyrie Irving (I'll say more about Irving's flat Earth conspiracy theory beliefs in Chapter 6 as well). It isn't clear just how much Hughes was a committed "flat Earther" or whether it was all just a pretense for his stunt, but his public endorsement of the belief—he had "Research Flat Earth" painted on an earlier version of his rocket ship—did help to raise the thousands of dollars required to make the event possible. In any case, his stated intent of witnessing the true shape of the Earth with his own eyes came to an abrupt end when his rocket ship struck part of the launch assembly on the way up, causing it to take a nosedive after a brief ascent. He died from the subsequent impact with the ground.[10]

If these two object lessons are, like Heaven's Gate, still too tempting to write off as tales of individual lunacy, consider the untold numbers who fall ill and die each year due to belief in medical misinformation. For example, according to the UK charitable organization Avert, some 21 million people were living with HIV in sub-Saharan Africa in 2019, with new cases representing 60% of all new HIV infections worldwide. Despite ongoing public health interventions and education campaigns, enduring folk beliefs about noninfectious causes of AIDS and false claims that it can be cured by witchcraft remain significant contributors to this overrepresentation. In 2019, the United States similarly experienced a mini-epidemic of nearly 1,300 cases of measles—an infectious disease that the US Centers for Disease Control and Prevention (CDC) declared to be eliminated from the United States back in 2000—across 31 states due to beliefs in misinformation about vaccines causing adverse effects in children.

As we know all too well now, "anti-vaccine" beliefs blossomed around the world during COVID-19 thanks in part to false claims that vaccines against the SARS-CoV-2 virus caused sterility or contained microchips that would be used for tracking purposes. Consequently, only about half the population of the United States and 40–66% of those in the European Union planned to get a COVID-19 vaccine through the spring of 2021.[11] By the summer of that year, the Associated Press was reporting that nearly all of the COVID-19–related deaths in the United States—occurring at a rate of about 300 per day—were among those that hadn't been vaccinated.[12] Subsequent studies revealed that while COVID-19 deaths became increasingly common among those who were vaccinated as the SARS-CoV-2 delta variant spread around the world, death was still 11–14 times more likely among the unvaccinated.[13] As the majority of the US population eventually embraced vaccination, boosters lagged behind new omicron variants, and unvaccinated people developed immunity through infection, that disparity narrowed by 2023, but researchers at Brown University estimated that more than 300,000 deaths from January 2021 through April 2022—nearly half the total COVID-19 deaths during that period—could have been prevented had vaccination been fully accepted early on.[14] Doing the math on the nearly 7 million deaths attributed to COVID-19 worldwide by the end of 2023, and recognizing that disbelief in that statistic and unsubstantiated counter-claims about vaccine-related deaths have become political dogma, it's hardly a stretch to expect that vaccine hesitancy due to belief in misinformation may very well end up being more lethal than any other misbelief in our lifetimes (I'll revisit anti-vaccine beliefs later in this chapter and the next as they relate to misinformation, as well as in Chapter 6 as they relate to conspiracy theories).

Whether we're talking about Heaven's Gate, Pizzagate, flat Earthers, or COVID-19, two conclusions should be clear. First, false beliefs can take root in individuals independently of delusional thinking or mental illness, and they can spread across entire populations in spite of education to the contrary. They can even arise in the absence of a reliance on naïve realism, a deficit of intelligence, or any faulty reasoning process per se. All it takes for any of us to embrace a false belief is that we fall for misinformation that we find not only on the internet, but all throughout the world where it's ubiquitous. In other words, as I suggested in the previous chapter's discussion of confirmation bias and echo chambers, all it really takes is to accept what someone else tells us. Second, belief in misinformation always carries at least the potential for harmful—and sometimes deadly—consequences. But if misinformation is so dangerous, why does it flourish? Why is there so much of it out there, and why do we find it so appealing? I'll explore answers to those questions throughout the rest of this chapter.

The Flea Market of Opinion

When I drove home late at night after being on call as a medical resident, I would occasionally stumble on *Coast to Coast AM* while scanning up and down the radio dial for something to keep me awake. I have to admit that I found the show entertaining—and perhaps informative as a budding psychiatrist—particularly when Art Bell would field middle-of-the-night calls from listeners to chat about the paranormal. When I was younger, I likewise enjoyed the occasional perusal of *The Weekly World News*, a notorious tabloid that, despite its self-billing as "the world's most reliable newspaper," regularly featured stories about space aliens, UFOs, and other outlandish topics written with an obviously tongue-in-cheek tone that few took seriously. And even now, when I'm waiting in the grocery store checkout line and not staring down at my smartphone, my eye is just as likely as anyone else's to be drawn to the headlines of tabloids like the *National Enquirer* that provide a steady source of titillating celebrity gossip.

At its peak popularity, when Bell was still its host, *Coast to Coast AM* was the most syndicated show on radio with more than 10 million listeners. *The Weekly World News* printed 1.2 million copies of its issues at the height of its success, and when *The National Enquirer* changed hands in 2019, it sold for $100 million. Which is to say that this kind of media—recognized by most of us as providing entertaining fiction rather than factual news—has provided a long-standing formula for financial success. Over the past several decades, however, coincident with the rise of cable TV and the internet, fiction masquerading as news is no longer confined to late-night AM radio and the grocery store checkout line as a form of entertainment. It's all around us, trying to pass itself off as the real thing. Meanwhile, few of us have ever been taught how to tell the difference.

Since I spent the previous chapter focused on the internet, I'll turn now to the example of political news coverage on television to back up this claim. A 2019 Pew poll found that television remains the primary source of political news in the United States, just narrowly edging out social media and other online sources across all age groups and still beating it out by a large margin for those older than 50.[15] Similar polling in recent years has revealed that, within television news that includes national network, local, and cable TV programming, cable now surpasses network TV as a primary news source, although the networks still maintain their long-standing dominance in terms of a larger overall audience.[16] This suggests that over the past several decades audiences have increasingly shifted from tuning into news "reporters" like Walter Cronkite, Dan Rather, Tom Brokaw, Peter Jennings, and Ted Koppel to listening to news "commentators" like Rush Limbaugh, Glenn Beck, Bill O'Reilly, Sean Hannity, Laura Ingraham, Megan Kelly, and Tucker Carlson on the political right and Rachel Maddow, Keith Olbermann, Anderson Cooper, and Chris Matthews on the left. These days, we've also become accustomed to taking in news-as-comedy courtesy of Bill Maher, Jon Stewart, Stephen Colbert, John Oliver, Trevor Noah, and Samantha Bee and even news-as-conspiracy theory by the likes of Alex Jones. By following news programming presented as attention-grabbing op-ed "infotainment" in the style of "political sportscasting,"[17] we're now increasingly likely to conflate objective news with our favorite media personality's subjective interpretation of the news. In other words, we risk shortchanging our chance to think through the issues of the day by looking at the facts in favor of allowing others to tell us how to think and feel about the world and each other.

Nowhere was this distinction better spotlighted than during an awkward 2016 interview on the *Fox News* show *The O'Reilly Factor* in which host Bill O'Reilly invited Ted Koppel, the ABC *Nightline* host who retired in 2005, to talk about how to interview Donald Trump, who was running for president at the time and had not yet been fully embraced by the Republican Party.

O'Reilly led off the exchange by asking, "Donald Trump—I've interviewed him a number of times—not an easy interview. How would you do it?"

Right off the cuff, Koppel shot back, "You know something Bill, you and I have talked about this general subject many times over the years—it's irrelevant how I would do it. And you know who made it irrelevant? You did. You have changed the television landscape over the past 20 years—you took it from being objective and dull to being subjective and entertaining and in this current climate, it doesn't matter what the interviewer asks him. Mr. Trump is going to say whatever he wants to say, as outrageous as it may be and the fact of the matter is his audience, as much as anything, is not even a television audience—it's an audience on Twitter. They deal in messages of 140 characters or less, which keeps it nice and simple."

O'Reilly then parried, asking, "OK, but, you know, your old network ABC does interview Mr. Trump on a regular basis, and ... you've got to come in with a strategy with him ... you've got to come in with sharp questions ... our job, whether I'm a commentator or a reporter is to get as much information ... and show the viewer who the person really is. So again, I'll go back to ... he's sitting on *Nightline* ... you're opposite him ... how do you do it?"

Koppel answered, "Well, the first way you do it is not in the interview—you do it by some reporting ... it's an old-fashioned concept."

Later during the exchange, O'Reilly pressed Koppel again, asking, "Would you as an anchorman ... you obviously don't like Donald Trump ... if you're in the chair now ... today, with all the things that you pointed out ... it's a whole different ballgame on cable TV ... commenters like me have just ruined the country; I cop to that ... I've ruined everything, although the journalists outnumber the commentators by about 50 to 1 and maybe the journalists aren't as powerful as they should be ... but anyway ... would you show your disdain at a certain point for a certain candidate ... under these new rules on television, is that allowed?"

"No," Koppel replied, "I don't like the new rules of television, and quite frankly, I don't think I would adhere to the new rules of television.... It's not a question of what I personally think—it's a question of whether there is any substance there ... you asked me what questions I would ask.... I think the first thing that has to be demonstrated is a little bit of journalism—go into some of the details of who and what Mr. Trump actually is, what those policies amount to, and then after you've laid it out ... you remember on *Nightline* that's what we used to do ... then you talk to the candidate and you say, 'Why is it that we don't have anything more just than fluff?'"

Pivoting, O'Reilly replied, "OK, but the problem is that now, the network news ... CBS, NBC, and ABC, proceed to be liberal ... so the people that you might want to persuade are saying, 'You know what? They don't like them and they're going to twist their reporting'—and it's the same thing with *The New York Times* and *The Washington Post*, that a lot of the straight reportage that used to be accepted is now questioned because of this perceived liberal bias. Correct?"

Koppel iterated, "And the fact of the matter is, Bill, you deserve both credit and you have to accept responsibility. For the past 20 years, as I said at the onset, you have been changing the landscape. The fact of the matter is—it's hard to believe these days—but 30 years ago, a television network anchor, Walter Cronkite, was the most trusted man in America. There's not a man today—yourself included—on television as an anchor who is trusted by anything approaching the majority of the American people."[18]

Much like Koppel, I waxed nostalgically in the previous chapter about growing up in a simpler time when there were only three major television networks that were relied upon to report the news. I'm not suggesting that this was necessarily better—the democratic diversity of voices that both report and comment on the news today, challenging mainstream dogma when it deserves to be challenged, should in theory represent an upgrade over not only the state-run news machinery of authoritarian countries, but even the trusted cadre of a select few older white men that I grew up with here in the United States. And yet, there has been a tradeoff. Within the limitless expanse of information at our disposal today, the unintended consequence of informational diversification has been that the line between reporting objective facts and expressing subjective opinion is often blurred beyond recognition. Tabloid journalism is everywhere, with few signposts to guide consumers how to readily distinguish reliable information from the misinformation that sits right alongside it. Not coincidentally, and just as Koppel claimed, trust in mass media has steadily eroded over the past half-century so that the majority of Americans today no longer trust what they read or hear from newspapers, television, or the radio.[19]

Koppel has been railing against what he sees as the sad state of journalism for well over a decade now. In 2010, he penned a *Washington Post* op-ed entitled "The Case Against News We Can Choose" that decried partisan political commentary on both Fox News and MSNBC and a "pervasive ethos that eschews facts in favor of an idealized reality."[20] During a 2015 interview on *The Late Show with Stephen Colbert*, he described modern news journalism as consisting of so many "fragmented" sources, each in competition with one another to capture our attention, that news programming has devolved from giving people "the news they need" to giving them "the news they want."[21] In a more recent 2019 interview at the Council on Foreign Relations, Koppel noted that the "marketplace of ideas" has become the "market place of opinions ... a marketplace of lying and calumny ... a marketplace in which, tragically, the most outrageous views get the greatest attention."[22] He denounced the "democratization of journalism"—where anyone and everyone is given a desk and microphone to bloviate about world events—as vastly overrated and even dangerous, pointing out that we wouldn't welcome the democratization of roofers or doctors, allowing those with no training or expertise to fix our houses or perform an appendectomy.

To some, Koppel no doubt comes off as an aging curmudgeon—his *Washington Post* op-ed was described as the pining lament of "the Grumpiest Old Man"[23] while another *Washington Post* op-ed referred to his exchange with O'Reilly as "gloomy" and "scolding"[24]—trampling on the democratic ideal of a free market of opinion. And yet the claim that the egalitarianism of a direct democracy in which everyone within a multitude is given equal voice can imperil the greater good isn't nearly as undemocratic or un-American as it might sound. James Madison and the other framers of the Constitution felt strongly that the "instability, injustice, and confusion" of factionalism arising from the free-for-all of "pure democracy" ought to be remedied by the formation of a Republic, whereby "public views" would be "refined and enlarged" by "passing them through the medium of a chosen body of citizens, whose wisdom may best discern the true interest of their country."[25] While the wisdom and true meaning of this premise remains actively debated today,[26] Koppel's nostalgia for the good old days when we had trusted "gatekeepers" like Walter Cronkite—that is, journalists who would investigate news stories and curate them for our consumption and who we could rely on for objective reporting—can be thought of as a modern-day reflection of this founding principle of representative democracy.

The great hope of the marketplace of ideas created by cable TV and the internet was that truth would rise to the top, but the evidence is all around us that the opposite has occurred. Without recognized gatekeepers serving as "librarians" to show us reliable information, the marketplace of ideas is much less a glass of fresh milk in which the cream rises and much more a polluted ocean where the flotsam and jetsam float up and coalesce on the surface as vast islands of garbage. Our current media landscape may allow a democratic diversity of opinions to be heard, but, as I discussed in the previous chapter, it's often not the meek or underrepresented voice that's elevated. Instead, attracting our attention amid a din of so many voices often necessitates being the loudest, the most salacious, the most outrageous, the most sensationalist, and the most untrue. The marketplace of ideas is bustling, but it bustles like a flea market, with all manner of informational salesmen hawking their wares, much of which is unreliable junk that's better left untouched.

Truth Decay, Misinformation, and Fake News

I've spent a fair amount of time thus far implicating the rise of cable TV and the internet in creating today's unbridled flea market of opinion, but something similar has occurred at other points in history before either existed. Nicholas Lemann, Dean Emeritus of the Columbia University Graduate School of Journalism, penned a 2006 op-ed that presaged Koppel's in debating the merits of "citizen journalism" versus content that's curated by gatekeepers or what he called "the priesthood" in the internet era.[27] He noted that our modern dilemma echoes that of late seventeenth- to early eighteenth-century England, when the advent of the printing press resulted in an explosion of pamphlets and periodicals that were "cheap, transportable, and easily accessible to people of all classes and political inclinations" and offered "a sort of interactive entertainment" written in a "distinctive, hot-tempered rhetorical style" that "delighted in mocking or even abusive criticism, in part because of the conventions of anonymity." Sound familiar?

RAND researchers Jennifer Kavanagh and Michael Rich have described similar cycles of "truth decay" occurring at other periods throughout American history following the emergence of other forms of new media.[28] Truth decay, they say, is characterized by (1) an increasing relative volume of opinion and resulting influence of opinion over fact, (2) a blurring of the line between opinion and fact, (3) declining trust in formerly respected sources of factual information, and (4) increasing disagreement about facts and analytical interpretations of data. Within their analysis, published in 2008, they found evidence of truth decay during the "yellow journalism" newspaper era of the 1880s and 1890s, after the advent of radio and tabloid news journalism in the 1920s and 1930s, and following the invention of television in the 1960s and 1970s. According to Kavanagh and Rich, the common threads running through these time periods, as well as during the modern era from the 2000s onward, have been the first two components of truth decay: an increase in published opinions over facts and a blurring of line between the two. Today's truth decay is therefore hardly novel and seems to reflect an ebb and flow cycle in response to innovations in media. But, unlike the previous three eras, Kavanagh and Rich found that all four components of truth decay can be detected amid our current crisis, with not only an admixture and conflation of facts and opinions, but also mistrust in previously respected sources of information and increasing disagreement over facts and interpretations of data. Curiously, while mistrust was also present in the 1920s and 1930s and prominent in the 1960s and 1970s, they found that disagreement over facts has been unprecedented during previous cycles of truth decay and is therefore unique to our time.

Why has the democratization of opinion occurring in the wake of cable TV and the internet resulted not only in the conflation of opinions and facts but also in an inability to agree on facts and truth? To answer this question, we should first acknowledge that while there's a crucial difference between objective reporting and opinion, the latter doesn't necessarily represent misinformation—that is, information that's wrong or false. In fact, opinions are sometimes spot on or later turn out to be true, regardless of whether they come from experts, which is often what makes them so tantalizing. Right or wrong, however, opinions can be defined as subjective interpretations of the available data or speculations based on insufficient evidence to elevate them to the level of objective facts. And yet they're rarely framed that way. How often do we hear news commentators say, "let me present two different perspectives on this complex topic"? How often do we hear them, or anyone else, say, "I could be wrong, but this is my take"?

With opinions declared with the same confident language as facts in today's media landscape, it's become all too easy to mistake misinformation for accurate news. The path to this unfortunate state of affairs is one that has, over time, been paved by a combination of a free-market economics and federal deregulation. In his *Washington Post* op-ed, Koppel highlighted that broadcast television news was a money loser for the major networks back in the day, so that it was only sustained through mandates set forth in the Radio Act of 1927 that required networks to serve "public interest, convenience, and necessity."[20] In much the same way, the "Fairness Doctrine," subsequently established in 1949 by the Federal Communications Commission (FCC), required networks holding broadcast licenses to present "controversial topics of public importance" in a manner that was fair and balanced, including a requirement to air contrasting viewpoints.[29] However, the Fairness Doctrine only applied to network programming, not cable TV, and, in response to lawsuits that claimed it violated free speech, it was revoked altogether in 1987.

With the removal of federal regulation of television news outside of sources like the Public Broadcasting Service (PBS) today, programming has transformed in kind. Modern news audiences rarely pay for subscriptions to individual news sources anymore—a mere 3% of the US public relies on print news as a primary source,[15] and with access to free programming provided over a perpetual 24-hour news cycle on cable TV and the internet, generating revenue now depends on attracting eyes and clicks to both programming and advertisement. Consequently, as if taking a page from the playbooks of *The National Enquirer* and *The Weekly World News*, media producers have increasingly attracted our eyes and attention-challenged brains with the irresistible pull of clickbait in the form of bullet points, soundbites, rumors, rants, and uber-opinionated partisan hot takes. And just as Art Bell hinted at in response to the Heaven's Gate suicides, journalistic integrity and responsibility often plays second fiddle to entertainment when trying to capture and sustain the attention of viewers. Within this modern model of unfettered profiteering, cable TV news revenue dollars now number in the billions each year, far surpassing those of network TV news,[16] and even relative nobodies can generate eight-figure incomes as social media influencers and YouTube channel hosts.[30]

* * *

A 2018 study by researchers at the Massachusetts Institute of Technology (MIT) goes a long way toward understanding why misinformation can be so profitable. It compared how news stories that were verified as true were shared and spread across Twitter compared to stories or rumors that were determined to be false.[31] Looking across 126,000 news stories and more than 4.5 million tweets from 2006 to 2017, the study found that "false news" (a term the authors prefer over "fake news") spread significantly faster, farther, and more broadly, reaching far more people than factual news. More specifically, Twitter users were 70% more likely to retweet false stories than true ones.[32] The difference persisted across information that pertained to politics, urban legends, business, terrorism and war, science and technology, entertainment, and natural disasters. Needless to say, this striking revelation has an important implication. If one wants to maximize an audience on social media—that is, to "go viral"—there may be no better way to achieve success than to peddle attention-grabbing headlines that aren't true. And there's no reason to suspect that one's success in this regard would be limited to Twitter, social media, or even the internet.

As a timely and particularly worrisome example of this marketing principle, the virality of misinformation has become a well-known challenge driving false beliefs that underlie vaccine hesitancy, as I suggested at the start of this chapter. Although the efficacy and safety of vaccines is well-established within the medical community, with a mountain of evidence published in the medical literature clearly demonstrating no greater incidence of autism among those vaccinated,[33] vaccine misinformation abounds outside of that confined and well-regulated informational space. For example, a 2010 study of Google searches using the term "vaccination" found that 71% of the results came from sources that were classified as anti-vaccination sites.[34] While social media platforms have taken steps to revise their algorithms to deprioritize such medical misinformation in recent years, sites that promote vaccine misinformation nevertheless still remain prominent and popular across Google, Facebook, and YouTube.[35] As of October 2019, anti-vaccine Facebook pages outnumbered pro-vaccine pages more than 2:1, with followers growing more rapidly and interacting with other groups with ideological overlap, such as those focused on "wellness" or more generalized safety concerns.[36] Meanwhile, followers of pro-vaccine groups were more static, without evidence of nearly as much interaction with or ideological conversion of vaccine "fence-sitters." In July 2021, a poll by *The Economist* and YouGov found that 17% of a nationally representative sample believed that vaccines "have been shown to cause autism" and 25% were unsure.[37] Twenty percent believed that "the US government is using the COVID-19 vaccine to microchip the population" and 14% were unsure. Another poll from the same time period conducted by the COVID States Project found that belief in COVID-19 misinformation was higher, and vaccination rates lower, among those who relied on Facebook, Fox News, or Newsmax for COVID-related news compared to those using other mainstream media sources.[38]

Such sobering statistics illustrate all too well that misinformation can often beat out reliable information within media spaces that aren't regulated by gatekeepers and other filters of reliability, with potentially harmful results. By the summer of 2021, a year and a half after the World Health Organization warned of a looming "Infodemic," the impact of false information about vaccines on COVID-19 mortality was enough for President Biden to declare that misinformation spreading on social media platforms like Facebook was "killing people."[39] And yet, thanks to such misinformation, a nontrivial proportion of the population now believes that it is vaccines that are doing the killing. It has become an inevitability that someone, somewhere on the internet will blame the illnesses or sudden deaths of celebrities on vaccines, as was done with Buffalo Bills football player Damar Hamlin; USC freshman and NBA hopeful Bronny James; comedian Bob Saget; singer Tina Turner; and actors Betty White, Lance Reddick, Ray Liotta, and Jamie Foxx.

The problem of medical misinformation is by no means confined to the narrow scope of vaccines. A systematic review of articles published on social media between 2012 and 2018 found that misinformation related to other medical topics including cancer and nutrition was also on the rise.[40] A 2021 study analyzed 200 articles posted on social media about cancer and found that 33% contained misinformation and 31% contained harmful misinformation.[41] These articles had significantly greater engagement than articles that were factual. It therefore seems clear that the metaphor of "going viral"—with misinformation about medical illnesses spreading as rampantly as an infectious disease or a malignant cancer—has become far more fitting than was likely intended when that phrase was first coined.

***

There's little doubt that the free-for-all landscape of digital media has become a veritable Petri dish where misinformation flourishes. But, as if that's not problematic enough, misinformation has also found a way to sneak into informational spaces that have been purposely designed to safeguard against it, such as scientific and medical research journals. To understand what has changed over the past few decades, let's first spend a moment discussing how the publication process is supposed to work.

When researchers submit manuscripts for publication to academic and scholarly journals, acceptance or denial is traditionally based on an editorial peer review process. During the first round of review, manuscripts are sometimes rejected off the bat, especially by top-tier journals that typically accept only the most impactful and rigorously conducted research. When papers are sent out for peer review, reviewers critique the work and often request that the authors address questions about their research and revise their manuscripts accordingly. Second and third rounds of review may follow, taking as much as a year or more for a publication decision to be made by the journal and for the paper to appear in print. When a manuscript is rejected, the authors may go back to the drawing board to correct problems in their study or their data analysis, and they might consider resubmitting their manuscript to another, often lower-tiered, journal. This can be a humbling process—something I've experienced firsthand on many occasions—though a considerable upside is that peer review amounts to free expert fact-checking and editorial advice that almost always results in a stronger end product and weeds out work that fails to meet quality standards. Since researchers aren't charged a submission or publication fee for this review, and since reviewers are blinded to a manuscript's authors and aren't paid for their work, conflicts of interest are also minimized.

While this publishing model still remains largely intact, the internet has transformed the academic publishing landscape just as cable TV and the internet have transformed news media, with an exponential expansion of options for researchers to publish their work. While scientific journals have historically depended on subscriptions—often by university and other research institution libraries—for revenue, this meant that access to published research was limited to a narrow and privileged audience. As a result, "open access" journals emerged in the 1990s as alternatives that were financially supported by publication fees paid by authors, with no subscriptions needed to read their content online. While this new model aimed at broadening access was well-intentioned, it generated an inherent conflict of interest whereby profits were increased by publishing more papers.[42] Consequently, while there are many reputable open-access publishers, there's been an explosion over the past couple decades of what Jeffrey Beall, a library scientist and former professor at University of Colorado Denver, labeled "predatory publishers" that exploit the open access model with a pay-to-print system that charges considerable fees to authors in exchange for less rigorous peer review and an easier path to publication.

Indeed, hardly a week goes by for me these days without receiving an email from a predatory journal soliciting a manuscript and occasionally asking me to join their editorial boards. This is not only flattering—especially if you're unaware of what's going on—but is further incentivized by an academic promotional system that demands that researchers "publish or perish" so that some have characterized the relationship between such journals and authors as more symbiotic than predatory.[43] Although it's been argued that the widespread proliferation of open-access publishing represents an "ideological good" in the same vein as the democratization of news reporting, it can likewise be difficult to distinguish reputable open access journals from predatory open access journals that deliberately mimic the titles of top-tier journals and even go so far as to list reputable academic scholars as editorial board members without their knowledge or consent. Beall has therefore argued that the net effect of open access publishing has been to detract from the overall quality of academic and scientific publishing to the point that "there is a journal willing to accept almost every article, as long as the author is willing to pay the fee,"[44] and pseudoscience and "junk science" can now be routinely found masquerading as the real thing.[45] While Beall and others have maintained lists of predatory publishers to help us identify them, the line between predatory journals that are deceptive and lower-tier journals that merely have shoddy publishing standards is blurry and continues to be debated.[43] So it is that naysayers of scientific consensus can now cite articles legitimizing the likes of a vaccine-thimerosal-autism link, Morgellons, and even "chem-trails" while being unaware of the reliability of such claims or the quality of the journals in which they appear.

In addition to predatory journals, a number of online repositories have been developed to facilitate the publication of "preprints"—research articles that haven't undergone peer review and haven't yet been accepted for publication. This is often appealing to the research community as a way of using the latest data to guide further investigation in real time, similar to the long-standing practice of presenting and discussing preliminary study results at research meetings. But, through online publication, preliminary research findings are now shared beyond the scientific community on a much larger scale under the cloak of authority. Although both policymakers and the general public are often eager to hear about cutting-edge research findings—such as during a global crisis like COVID-19—the reliability of such findings should be regarded as tenuous. After all, there's no guarantee that a paper published on a preprint server will ever make it through peer review and actually be published in a legitimate medical journal. And yet, the findings are often cited and publicized by the news media, social media influencers, and politicians nonetheless.

When peer review is short-shrifted or bypassed altogether, the end result isn't only the potential spread of misinformation but also a weakening of trust in and the perceived reliability of science as an institution of authority. Predatory journals and preprint publications may be a big reason that, as Kavanagh and Rich found in their study of truth decay, we now find ourselves in an unprecedented era when there's both mistrust in previously respected sources of factual information and widespread disagreement about facts and interpretations of data. But, despite what some have claimed,[46] it's not science that's creating the crisis so much as the erosion of scientific publishing standards within the much larger, unregulated flea market of opinion.

To be clear, it's not as if the integrity of journalism and scientific publishing is dead. There's plenty of reliable, objective news reporting out there, just as there's plenty of reliable, objective scientific and medical research being published. Rather, our modern dilemma lies in how to navigate the flea market of opinion to find the good stuff and, given how human brains search for information with confirmation bias, to what extent that's possible and to what extent we can agree on what the good stuff is anymore.

Motivated Reasoning and Identity Protection

After his clash with Bill O'Reilly, Ted Koppel would go on to spar with O'Reilly's Fox News colleague Sean Hannity, deriding him as "bad for America."[47] In response, Hannity fired back, "We have to give some credit to the American people that they are somewhat intelligent and that they know the difference between an opinion show and a news show.... You're selling the American people short."

So, who's right here? Is Koppel selling the American people short, or is Hannity giving us too much credit? Two recent legal cases set the stage for an answer. In 2019, MSNBC commentator Rachel Maddow was sued for defamation by the One America News Network after she claimed that it was "literally paid by Russian propaganda."[48] The case was later dismissed by a US district judge who ruled that Maddow's "exaggeration of the facts ... was consistent with her tone up to that point" and that a "reasonable viewer would not take [her opinion] as factual." The next year, Fox News commentator Tucker Carlson was similarly sued for slander by Karen McDougal, who Carlson claimed was extorting President Trump with allegations of an affair. Following the arguments presented by Carlson's lawyers, the judge in his case ruled that the "'general tenor' of [Carlson's] show should ... inform a viewer that [he] is not 'stating actual facts' ... and is instead engaging in 'exaggeration' and 'non-literal commentary.'"[49] "Given Mr. Carlson's reputation," the judge wrote, "any reasonable viewer 'arrive[s] with an appropriate amount of skepticism' about the statements[s] he makes."

We can see then that Maddow's and Carlson's successful defenses both hinged on their lawyers' admission that they're in the business of sensationalism and hyperbole, not the objective reporting of facts. However, while the judges ruled on the side of Hannity's charitable view of audiences' ability to tell the difference, the evidence from research to date is more supportive of Koppel's concerns and those of Kavanagh and Rich about truth decay. We, the American people—as well as our fellow human beings across the world—aren't particularly good at separating opinion from news or false news from reliable news. By way of evidence, researchers at the Stanford History Education Group presented middle school, high school, and college students with examples of various forms of digital media including news articles, social media posts, advertisements, photographs, and videos and asked them to rate the content based on reliability.[50] At all levels of education, students struggled with tasks like distinguishing real news sources from "native advertising" designed to promote or sell a product under the guise of a news story, questioning the source of random photographs accompanying informational claims, and detecting the potential political bias of poll data. And, lest we chalk these findings up to the ignorance of youth, other research has demonstrated evidence of a "gray digital divide" in which it's older people who are more naïve to the world of online information and most likely to share false news on social media.[51] The bottom line, as I claimed earlier, is that few of us have had the requisite education—or even *any* education—in media literacy that would allow us to easily distinguish reliable information from false information or real news from "fake news" across the vast number of informational sources that are collocated within a common space. Fake news may be relatively easy to spot when we see the likes of *The Weekly World News* in the grocery store checkout line, but when it's offered right alongside the real thing, discernment is a major challenge that few of us are well-equipped to overcome.

As Koppel suggested, one reason that we've become so vulnerable to misinformation at the expense of agreeing on facts is that, in today's media landscape, consumers are now fed the information we *want* rather that the information we *need*. Within the flea market of opinion, there's something for everyone, and, as I discussed in Chapter 3, confirmation bias on steroids means that we can now rapidly locate information that matches the convenient untruths to which we hold tight and allows us to deny the inconvenient truths that we prefer to ignore. Objective evidence has taken a back seat to what the comedian Stephen Colbert referred to as "truthiness," a term that the Merriam Webster Dictionary declared the Word of the Year in 2006, based on Colbert's description of it as "truth coming from the gut, not books" and the American Dialectic Society's definition as "the quality of preferring concepts or facts one wishes to be true, rather than concepts of facts known to be true."[52] Guided by the intuition and emotion of truthiness, we allow ourselves to be misled by the self-satisfaction of confirmation bias so that we often prefer false news over reliable news, preliminary results over peer-reviewed results, and opinions and "alternative facts" over actual facts. By way of analogy, what we're fed and what we consume in terms of news and information these days often mirrors what we're fed when we shop in the supermarket or eat out—a plethora of processed fast foods laden with the empty calories of high fructose corn syrup that appeal to our taste buds but detract from our overall physical and mental health.

Another significant and related driver of our unhealthy appetite for misinformation is something called "motivated reasoning," which Koppel alluded to when he told Hannity that modern fans of politically biased news prefer ideology over facts.[47] From a psychological perspective, although motivated reasoning is often equated or lumped together with confirmation bias, they can be understood as distinct cognitive processes. Thinking back to Daniel Kahneman's proposal of two modes of decisional thinking that I discussed in Chapter 2, confirmation bias can be modeled as more of an intuitive and emotional kind of fast thinking that often favors "truthiness" over truth—it directs our attention toward information that supports our pre-existing beliefs and away from information that refutes them. Motivated reasoning takes things a step further as a more rational and deliberate process that governs how we digest and interpret that information to justify what we already believe as well as to conform with what we think we should believe based on the ideological groups to which we belong.[53] It can therefore be thought of as involving social or cultural cognition, depending not only on the interactions between information and our brains, but also on the triangulation of information, our brains, and our ideological affiliations.

Instead of trying to dispassionately gauge the veracity of the information that we encounter or seek out, motivated reasoning entails analyzing it so that it fits into our existing worldviews and those of our affiliated social groups. So, if we're reading about evidence from a research study or listening to an interview with an expert that supports what we want to believe, we're likely to rate the quality of that study or the knowledge of that expert highly. And when studies and experts contradict our beliefs, we're likely to conclude that the study is flawed or that the expert is biased or not really an expert at all. Put more simply, we judge the evidence to be "good" when it supports our worldviews and "bad" when it doesn't, regardless of the actual quality of that evidence (you might recall that this is precisely what Lord, Ross, and Lepper observed in their 1978 experiment that I discussed in the previous chapter, so that the kind of post hoc analysis they described could be considered more akin to motivated reasoning than confirmation bias). Discounting evidence in this way has been variably called "disconfirmation bias," "motivated disbelief," "motivated skepticism," "motivated denial," and "motivated ignorance."[54] Ultimately, motivated reasoning amounts to dissecting and interpreting relevant facts and data in the service of safeguarding our identities based on our beliefs and affiliations with ideological groups. Consequently, this model of motivated reasoning is sometimes described by researchers like Yale University psychology and law professor Dan Kahan as "identity protective cognition" or by New York University professor of psychology and neuroscience Jay Van Bavel as an "identity-based model of belief."[55] I'll have more to say about the pitfalls of using our ideological beliefs to define our identities in Chapter 9.

Among the most easily recognizable and researched examples of motivated reasoning are beliefs that are determined by our political identities. These days, identifying as a Republican or Democrat, or as a liberal or conservative, can often reliably predict where we stand not only on political issues like abortion, gun control, immigration, voting rights, and foreign policy, but also on beliefs related to topics of scientific inquiry such as COVID-19 or climate change. This predictability occurs because of how confirmation bias and motivated reasoning interact with today's flea market of opinions, feeding off each other in a reciprocal fashion. We justify and defend our individual and group political beliefs by parroting talking points that we hear within our social media echo chambers and from our favorite partisan news commentators. Meanwhile, the commentators and social media influencers shift their own personas and ideologies in turn according to what partisan narratives are in vogue or are otherwise demanded by their audiences. Indeed, in early 2023, legal briefs revealed that after Fox News called Arizona in favor of Joe Biden during the 2022 Presidential election, panic over losing their audience share to the likes of Newsmax led hosts like Tucker Carlson and Laura Ingraham to subsequently promote unsubstantiated claims of election fraud despite deriding such claims in private communications.[56] Such pandering-for-profit over the reporting of verified facts has fueled a level of belief in misinformation and resulting political polarization—or "attitude polarization" as Lord, Ross, and Lepper called it—that, while hardly unprecedented here in the

United States, hasn't been seen in 50 years (I'll revisit the dilemma of political polarization more thoroughly in Chapter 8).

In 1990, Princeton University psychologist Ziva Kunda was one of the first to write about motivated reasoning, describing it as an "attempt to be rational and to construct a justification of [one's] desired conclusion" while detailing that

> People do not seem to be at liberty to conclude whatever they want to conclude merely because they want to.... They draw the desired conclusion only if they can muster up the evidence necessary to support it.[57]

And yet, because the flea market of opinion now features a plethora of conflicting information and misinformation side by side, the "evidence" to support any belief is all around us, as I discussed in Chapter 3. We can therefore avoid any conflict between beliefs and facts by shamelessly selecting misinformation through confirmation bias and motivated reasoning while deriding our ideological opponents as the ones who are misinformed so that we maintain what Kunda called an "illusion of objectivity." In short, instead of using facts to inform our beliefs, we use information—regardless of its veracity—to justify the beliefs that we and our affiliated social groups already have. As a result, scientific experts and even scientific consensus can be dismissed out of hand, just as fact-checking sites, guides that objectively rate media sources according to reliability and political bias, and lists of predatory journals can be discounted as biased themselves. Accordingly, the terms "misinformation" and "fake news" have now been hijacked and weaponized as euphemisms to describe whatever refutes our own beliefs and supports those of our ideological opposites, regardless of whether it's false or not.[58] With each side of almost any debate able to cite its own "evidence," we've come to a point where—as Kavanagh and Rich concluded—we're no longer willing to agree on something as seemingly fundamental as what counts as evidence, facts, or truth anymore.

Cognitive Laziness Versus Motivated Accuracy

It has been claimed that America has been imperiled by a wave of "anti-intellectualism" that has swept through the country over the past several decades—Isaac Asimov lamented a rising "cult of ignorance" back in 1980, and Carl Sagan warned of an evolving "celebration of ignorance" and a "dumbing down of America" in 1995.[59] However, we would do well to avoid the trap of accounting for where we are today by conflating motivated reasoning with an intellectual deficit. As I mentioned in Chapter 2, our vulnerability to cognitive biases isn't correlated with intelligence, and this has been specifically demonstrated for confirmation bias as well.[60] This means that both Koppel and Hannity are correct to a degree—we're not particularly good at separating facts from opinion and reliable information from misinformation as Koppel laments, but Hannity is right in that it's not a problem of low intelligence.

Not only does intelligence fail to protect us from belief in misinformation, it can make us more vulnerable to it under certain circumstances. Across several studies, Kahan has demonstrated that analytic thinking and "numeracy"—the ability to make sense of and reason through quantitative data—is sometimes associated with a greater tendency to employ motivated reasoning in the service of safeguarding beliefs.[61] This suggests that analytic thinking and the ability to interpret data arm us with the skills to defend our beliefs—to justify to ourselves and to others that we're right—even when they fly in the face of objective facts. This disappointing conclusion matches my experiences in high school debate class that I described in Chapter 3 and is reminiscent of how I ended Chapter 2, conceding that even experts aren't immune to unwarranted overconfidence in their beliefs. In much the same way, being intelligent—or, in this case, being adept at data reasoning—falls far short of guaranteeing immunity to false belief through motivated reasoning.

Fortunately, research on susceptibility to belief in "partisan fake news" paints a less discouraging picture. A series of experiments by University of Regina psychology professor Gordon Pennycook, MIT professor of brain and cognitive sciences David Rand, and their colleagues have shown that one's ability and propensity to engage in analytic thinking is indeed associated with a greater ability to distinguish between false and accurate news, as we would hope, even when false news aligns with our political ideology.[62] Pennycook and Rand argue that, contrary to Kahan's conclusion, belief in false news is therefore less about motivated reasoning or what we want to believe and more about a kind of "cognitive laziness" that can be overcome with concerted effort to think rationally. Put another way, Pennycook and Rand argue that their research supports Daniel Kahneman's "classical reasoning" model in which deliberate rational analysis helps us overcome the biased intuitions of heuristics.

The disparate findings of Kahan and those of Pennycook and Rand are most likely products of different experimental study designs along with the complexity of interactions that occur between confirmation bias based on prior beliefs and motivational reasoning processes based on partisan affiliations. In other words, our ability to steer clear of motivated reasoning might depend on a number of factors including our individual ability to think rationally, how we go about doing so, and our preexisting beliefs and ideological affiliations, as well as the specific topic or news headline that we're trying ascertain for truth. For example, it may be that our ability to think analytically might be compromised around certain complex topics like climate change (a frequent subject of Kahan's experiments) or around subject matter that's emotionally charged and therefore more likely to provoke feelings of anger or defensiveness.

It's also likely that, when participating in a psychology experiment, our ability to discern truth depends on what exactly we're being asked to do. If research subjects are given "party cues" that amount to telling subjects what information is endorsed by their political party, they might be more likely to give partisan responses suggesting that they're engaging in motivated reasoning.[63] But when subjects are explicitly asked to distinguish real and fake news, this may amount to a cue or warning that they might be presented with false information. Armed with the knowledge that we might be deceived, we may be more likely to activate our "bullshit detectors," so to speak, allowing us to slow down and think more skeptically and analytically, even when we encounter information that reinforces what we already believe or want to believe.

This explanation harkens back to Kunda's original characterization of motivational reasoning as consisting of two separable motives: the motive to be "accurate" and the motive to be "directional" or aligned with our group identities.[57] When given directional cues, we might be more likely to engage in motivational reasoning, but, when given accuracy cues, we might be more likely to engage in analytical thinking. There's good evidence to support this: for example, some research studies have incentivized the motivation to be accurate not only through cues, but also by providing monetary rewards (i.e., paying people for correct answers).[64] With such motivation in place, accuracy tends to win out over motivated reasoning, with research subjects also being more likely to admit their ignorance—to say they "don't know." That brings us back again to the conclusion of Chapter 2, where taking the time to think analytically allows us to scale back the unwarranted confidence in our unfounded beliefs and the appeal of misinformation that we encounter in the world. I'll return to the merits of analytical thinking in Chapter 10.

In the meantime, this all suggests that, in the real world outside of the psychology lab, we could benefit from acknowledging that misinformation is everywhere, with accurate information and misinformation sitting side by side within our modern media landscape. We would then do well to train ourselves to be cautious of news that we want to hear, especially when it reinforces our existing ideological beliefs and those of our ideological groups. Although "do your own research" has become a rallying cry for navigating this informational space unfettered by gatekeepers, if we go about trying to find support for what we already believe or want to believe and trying to win ideological arguments, we will more than likely fall victim to false beliefs through confirmation bias and motivated reasoning. That might make us feel better in the short term, reducing any perceived mismatch between our beliefs and facts, increasing self-satisfaction, and "owning" our ideological opponents, but it's not really "doing research." Whether it's a matter of cognitive laziness or motivational reasoning, the end result is self-deception—choosing to believe things that aren't true. That's not a mentally healthy way to believe, whether we're talking about ourselves or society at large.

5
The Disinformation Industrial Complex

> O, what a tangled web we weave, when first we practise to deceive.
> —Sir Walter Scott

Distributed Responsibility, Distributed Liability

A few years ago, Robert Drummond and his wife Margo, two middle-aged grandparents from the Midwest, found themselves in some serious hot water with the Internal Revenue Service (IRS). Facing charges of tax evasion, mail fraud, and wire fraud, there was a very strong possibility that they were going to spend the rest of their twilight years in federal prison. That they had not paid income tax for at least a decade wasn't in dispute, nor was the allegation that they'd assisted many others with efforts to go "free and clear," eliminating their financial debts by doing what they'd done—not paying taxes, filing liens on other people's property, and sending a barrage of legal documents to the IRS through the years that they believed would allow them to legally exempt themselves from government authority. What was in question was whether the Drummonds had "willfully" defied the law or whether they'd genuinely believed in "good faith" that what they were doing was within the bounds of legality. If the latter could be established at their trial, they might be able to avoid conviction based on the so-called *Cheek defense* that applies to tax law, which allows a rare exception to the rule that "ignorance of the law is no excuse."

Because of my academic work and writing on delusion-like beliefs, the Drummonds' defense attorneys reached out to me as an expert witness to determine whether mental illness might account for the couple's unusual beliefs and to assist with evaluating the viability of the Cheek defense in their case. Recalling the distinction between delusions and delusion-like beliefs that I discussed in Chapters 1 and 3, the key questions that I set out to answer were what the Drummonds actually believed, how strongly they believed it, and how they'd come to arrive at their beliefs in the first place. While interviewing the couple over the course of an afternoon, it became clear to me that they genuinely believed they possessed privileged knowledge about how to legally avoid paying taxes. They held this belief with high conviction, such that they were confident that they'd done nothing wrong and would be found innocent of the charges. Admittedly though, the more detailed rationale behind their confidence was both strange and convoluted, centering around a revisionist version of American history claiming that, after abandoning the gold standard in 1933, the

US government sold out its citizens as collateral for loans to other countries and replaced "common law" with "admiralty law" intended to regulate international commerce. They believed that by refusing to participate in standard practices such as getting a driver's license, filing a litany of documents to various government agencies written in a kind of nonsensical and idiosyncratically punctuated legalese, and suing the IRS, they could reclaim their status as "sovereign citizens," exempting them from admiralty law and even entitling them to some $630,000 held within a corporate trust established for all American citizens at the time of our birth.

Despite their rather odd—and patently false—beliefs, it was obvious that the Drummonds weren't delusional or otherwise mentally ill. For one thing, Mrs. Drummond was actually agnostic about the whole thing, deferring to her husband's expertise as she believed a good wife should. Mr. Drummond's beliefs were false and had something of a self-referential component, but were ultimately beliefs about the world rather than himself—he believed that all American citizens were entitled to or exempt from the same things he was, if they just followed the procedures that he'd followed himself. More to the point, the revisionist American history and claims about not being obligated to pay taxes that he espoused weren't idiosyncratic to Mr. Drummond: they were readily identifiable as the core dogma of a loosely affiliated, but growing group of people who self-identify as sovereign citizens.[1]

Like the Drummonds, sovereign citizens represent a subcategory of "tax protesters" and "tax deniers" who often run afoul of the law not only by tax evasion and filing false returns, but also by driving without a license, filing bogus liens for profit or as retaliation against perceived enemies, and engaging in "paper terrorism" by submitting a never-ending stream of fraudulent documents containing legalistic gibberish to the IRS and other government agencies in an attempt to legally extricate themselves from debt, bankruptcy, and property foreclosure. According to the Southern Poverty Law Center, sovereign citizens—who range from "hardcore believers" to those who are at least willing to test out some of the standard practices for evading the law—number in the hundreds of thousands here in the United States.

Robert Drummond had come to accept sovereign citizen dogma as gospel over the course of several decades after first hearing about it on a radio talk-show and subsequently reading about it on the internet and in published books with titles like *How to Avoid Paying Taxes Legally*. Eventually, he spent thousands of dollars attending seminars by self-proclaimed gurus of the sovereign citizen movement who laid out step-by-step guidelines for how to stop paying taxes, become debt-free, and reclaim one's sovereign rights. Once he mastered the formula, he set about trying to "pay it forward," assisting others to free themselves from debt and filing legal documents to that end on their behalf. Through this new calling, he managed to amass over a million dollars in personal profits from his clients, all the while feeling proud that he was helping others just as he'd been helped.

Despite my testimony in their federal trial, the Drummonds were found guilty of defrauding both the federal government and the people who'd paid for their assistance so that they were sentenced to prison, where they remain to this day. Neither I nor their defense attorneys were particularly surprised by the outcome—after all, many of the sovereign citizen gurus who'd instructed the Drummonds along the way had been similarly convicted and imprisoned in years prior, and I have yet to hear of a successful Cheek defense in a sovereign citizen case. But their situation did raise nagging questions in my mind about the many causal links in the chain of misinformation that resulted in the Drummond's beliefs, actions, and eventual fate. If the Drummonds were guilty of defrauding others despite their "good faith," what of the people who'd convinced them to buy into sovereign citizen dogma? Did those people genuinely believe it too, or were some of them just con artists running a scam by exploiting the desperation of people facing bankruptcy or losing their homes? Did Margo Drummond, who was mostly guilty of blind and unwarranted deference to her husband, really deserve to go to prison for doing what she believed made her a "good wife?" With so many links along the way in the chain of belief and action, who was ultimately responsible, and who should be held accountable?

Many of us did foolhardy things due to peer pressure we were children, prompting our parents to ask us, "If your friends told you to jump off a bridge, would you do it?" That rhetorical question is just as applicable to the Drummonds and the rest of us as adults, though its answer has become more complicated in the modern world where misinformation abounds and people are giving us bad advice about what to believe or not believe and what to do or not do all the time. When we act on misinformation that's given to us not just by friends and casual acquaintances, but news commentators, teachers, authors, social media influencers, celebrities, self-ordained gurus, and even politicians, shouldn't they share some of the blame for our actions? Wasn't Art Bell at least somewhat responsible for the deaths of the 39 members of Heaven's Gate due to posting doctored images claiming to show a UFO in the tail of the Hale-Bopp comet? Shouldn't conservative media personalities who spread the Pizzagate conspiracy theory bear some responsibility for James Maddison Welch? Shouldn't political leaders and news commentators who promoted the idea that Biden stole the presidency through a fraudulent election be held accountable for the deaths resulting from the violence at the US Capitol in January 2021? Shouldn't internet sites and social media platforms that facilitate the spread of such misinformation share some of the responsibility for its effects?

My answer to these timely questions of our day is "yes, all of the above." Although we should always take responsibility for our own actions and hold the perpetrators of crime accountable for theirs, we should also acknowledge and embrace the concept of "distributed responsibility," especially if we're trying to understand the cause of false beliefs and prevent people from acting on misinformation. And even within the black-and-white world of retributivist justice, where criminal sentencing is often rooted in punishment and revenge, we should consider moving closer to the concept of "distributed liability," especially in legal cases involving those with vested interests who stand to profit when other people act based on false beliefs.[2]

Mistrust and Misinformation

Back in Chapter 1, I noted that human beings aren't born thinking like scientists. As a result, we come to believe things for a myriad of reasons beyond weighing objective evidence, including our gut instincts, intuitions, and subjective experiences as well as relying on the testimony of others. In the previous chapter, I made clear that while naïve realism, confirmation bias, and motivated reasoning are all important drivers of belief, all it really takes to embrace a false belief is to accept what someone else tells us. The concept of distributed liability highlights that it takes two to tango—in other words, false beliefs often involve a dance with our brains following the lead of misinformation. In this chapter, following a discussion of trust and mistrust as drivers of belief and disbelief, I'll shift the spotlight away from our brains on the receiving end of that exchange to the misinformation that's on the production end—what University of Michigan psychologist Colleen Seifert describes as the more timely problem of "misinformation in the world" as opposed to only worrying about "misinformation in the head."[3] I'll also be sharpening the focus from the general topic of misinformation to the more specific problem of disinformation—that is, from information that happens to be untrue to information that's deliberately created and spread for its false counter-narrative. Since it's often difficult to judge intent or quantify just how much someone who tells us something actually believes it themselves, the line that separates misinformation and "false news" from disinformation and "fake news" is often blurred. Nonetheless, it's a crucial distinction when it comes to what we believe, amounting to the difference between believing things that are merely wrong and believing lies.

On the one hand, because none of us has enough subjective experience or personal expertise to possess knowledge on every subject, we must rely on what other people tell us in order to learn about the world. But, on the other hand, since we know that people can be wrong and that people do sometimes lie, we have to also maintain a kind of healthy skepticism that epistemologists, or philosophers that study knowledge, call *epistemic vigilance*.[4] Deciding who to trust or mistrust can be modeled as an assessment of "source credibility" based on the perceived trustworthiness and expertise of those supplying us with information.[5] As with all perceptions however, its accuracy depends on both the characteristics of the object in question—that is, the actual trustworthiness and expertise of an informational source—as well as certain characteristics of those doing the perceiving. While confirmation bias and motivated reasoning hamper our ability to objectively appraise whether news and information is true and accurate or false and fake, trust and mistrust are a big part of what determines just where those cognitive processes take us. Each of us has our own general capacity for epistemic trust or epistemic mistrust of both information and informational sources that can be conceptualized along a continuum with extreme gullibility at one end—where we believe everything we're told—and paranoid denialism at the other—where we trust no one and reject anything we don't want to believe out of hand. At either extreme, we become dangerously vulnerable to disinformation. When we're too gullible, we risk believing misinformation and disinformation alike simply because it's there. When we're too mistrustful, we risk getting drawn into disinformation counter-narratives that are designed to exploit epistemic mistrust, not only by spreading falsehoods, but also by further working to undermine the credibility of institutions of "epistemic authority"—that is, traditionally trusted sources of knowledge.

From a societal standpoint, there's good evidence that our current vulnerability to disinformation is more a result of mistrust than gullibility, ignorance, or stupidity as is often claimed. Recall that, in the previous chapter, I mentioned that trust in mass media has been in decline for some 50 years now. Indeed, a Gallup poll reported that about 70% of Americans had a "great deal" or "fair amount" of trust in newspapers, TV, and radio in terms of "reporting the news fully, accurately, and fairly" when polled in the 1970s, but, by 2016, that figure had dipped to just 32%.[6] In much the same way, about 75% of the public trusted "the government in Washington to do what is right just about always or most of the time" back in the 1960s, with that figure falling to 15% at the end of the Bush administration in 2011 and failing to rise above 25% through the Obama, Trump, and Biden administrations since.[7]

Why is it that we've become so wary of previously trusted institutions of epistemic authority like the news media and government? I'll have more to say about the roots of epistemic mistrust in the following chapter about conspiracy theories, but, as I explained in Chapter 4, declining trust in the media must first be understood as taking place within a modern informational landscape in which we're constantly presented with conflicting versions of the truth. No doubt, mistrust can also be traced back to instances of corruption, the exploitation of public office for personal gain, and the prioritization of special interests over the public good perpetrated by elected leaders and other government officials alike. However, our degree of mistrust seems to have grown far out of proportion to the actual number of trust violations that have occurred over that same period. Therefore, we should instead consider the possibility that we've been sold on a false narrative that traditional institutions of epistemic authority like the mainstream media or government agencies are no longer worthy of our trust and that we shouldn't believe anything we're told anymore.

In the previous chapter, I cited research demonstrating that false news travels farther and faster than accurate news and made the case for misinformation being potentially big business for cable TV networks, social media platforms, and the likes of predatory scientific journals as a result. As Jennifer Kavanagh and Michael Rich's work on truth decay suggests, the parallel growth of mistrust and misinformation isn't coincidental: the two factors are causally and reciprocally related, with misinformation filling the void left behind by epistemic mistrust and fueling more mistrust in the process. Put more simply, mistrust—whether it's earned or manufactured—sets the stage for us to be taken in by misinformation and deliberate disinformation in what can amount to a vicious cycle. Throughout the rest of this chapter, I'll be making the case that, these days, we're often told that we're being lied to by traditional institutions of epistemic authority, while those doing the telling are themselves operating in the service of trying to sell us on something else that isn't true.

The Apex Predators of the Disinformation Food Chain

In the previous chapter, I likened America's appetite for misinformation to an unhealthy appetite for a junk food diet. Here, I'll extend the metaphor by further clarifying the difference between misinformation and disinformation within what is sometimes referred to as the "disinformation ecosystem" or what I call the "disinformation food chain." Within the disinformation food chain, "apex predators" who are part of a "disinformation industrial complex" sit at the top, fomenting mistrust in institutions of epistemic authority and creating false information based on some underlying motive, usually to obtain either financial profit or political power. Daniel Kahan uses the term "opportunistic misinformers" to describe the apex predators who exploit "economic opportunism" to supply us with the information we want in the service of our motivated reasoning needs.[8] Others have referred to "conspiracy entrepreneurs"[9] who capitalize on the mass appeal of "disinformation porn." In the middle of the food chain are the "mesopredators" or "prosumers" who both consume disinformation and pass it on as misinformation while also creating novel disinformation themselves.[10] Finally, the "prey" or "pure consumers" lie at the bottom of the food chain as passive receptacles of misinformation and disinformation alike, sharing it with other consumers.

Within this model, Margo Drummond and the other people who did as her husband instructed were the prey as pure consumers of disinformation, taking it on faith. Most of us within the general public likewise spend most of our time at the bottom of the food chain as individual casualties of the structural harm that disinformation wreaks on a functioning democracy. People like Robert Drummond represent the prosumers who are both predator and prey. In the next few sections, rather than focusing on the victims of the disinformation food chain, I'll highlight a few key examples of apex predators working from within the disinformation industrial complex who profit from that victimization.

A War on Information

In 1996, Rupert Murdoch launched the Fox News Channel to counter the perceived liberal bias of the "big three" networks CBS, NBC, and ABC with a deliberately conservative perspective on the news. Since then, Fox News has become a go-to source of news for conservatives and has gained a firm foothold more broadly within mainstream American television media with its news commentary shows regularly sitting atop the list of most-watched cable TV news programs. As I mentioned in Chapter 4, however, Fox News was one of the first networks to declare Joe Biden as the new president of the United States, defeating the incumbent Donald Trump in the November 2020 election. As a result, and at Trump's urging, a sizeable number of viewers left Fox News in favor of the little-known cable network Newsmax so that, by December, it had, for the first time ever, scored some ratings victories over Fox within certain narrow demographics and programming time slots. By offering "a safe space in which Biden was not called president-elect and Trump was not yet defeated,"[11] Newsmax and the similarly hyperconservative One America News Network (OANN) were able to garner a significant bump in viewership by capitalizing on a rare opportunity in which conservative viewers were no longer satisfied with what Fox News was saying. Meanwhile, Trump mused about starting his own "Trump TV" channel, seemingly based on the similarly profitable model of offering disgruntled and mistrusting viewers what they want: disinformation in place of reality.

Selling fantasies and falsehoods as truth to a disgruntled audience is a formula for success that Alex Jones has mastered since launching his InfoWars website—presumably referring to a "war on information"—in 1999, just three years after Fox News came to be. Through the years, he has created a blustery and cartoonishly over-the-top persona that rants about conspiracy theories like Pizzagate or the baseless and outrageous claim that the 2012 shootings at Sandy Hook Elementary School resulting in the deaths of 20 children and 6 adults were a "false flag" operation staged by "crisis actors." In doing so, he has managed to attain both celebrity status and financial success—various news sources have reported that Jones amassed revenues of $20 million in 2014 and $165 million between 2015 and 2018, generated in large part through the sale of branded nutritional supplements bearing names like Super Male Virility and Brain Force Plus advertised on his InfoWars website.[12-14] In 2017, however, during a court hearing for a heated battle with his wife over the custody of their children in which his mental stability was called into question, his lawyer claimed that Jones was merely a "performance artist" who was "playing a character."[14] During a subsequent lawsuit by grieving Sandy Hook parents who were accused of having fictitious children and harassed after Jones made their names and addresses public, he backtracked on his claims in a recorded deposition, explaining away his false beliefs about the tragedy as follows:

> I, myself, have almost had like a form of psychosis back in the past where I basically thought everything was staged, even though I'm now learning a lot of times things aren't staged.... So I think as a pundit, someone giving an opinion, that, you know, my opinions have been wrong.[15]

Later, during an episode of InfoWars with his attorney on as a guest, he offered his audience a more measured account of what he meant during his "so-called deposition."

> Now ... people are probably asking, "Alex you're all over the news. They're saying all these incredible things about you. Why aren't you countering the fact that

you're one of the top stories in the country where you, quote, admit that you're mentally ill, that everything you say is BS?" Never said any of that. Didn't say that. That's a lie....

[W]e see claims that I'm saying ... that I have a form of psychosis. I wasn't diagnosed with that, I don't have a degree in that. What I was saying is that when the media lies so much and you've been told lies, it induces a state historically where people start believing nothing....

So, you've got people on one end that believe the Earth is flat and that the moon is made of cheese because the government says it isn't, and then you've got the other end where they say that President Trump is a Russian agent, we've got all the proof, and then Jake Tapper comes out and says we never said that today. So, you've got two ends, one lying on purpose, confusing people, the other not knowing what's real anymore and kind of floating around.

I was simply saying, they are inducing, instead of the Stockholm syndrome, where I just believe anything they say, I went the other way and said, "I don't believe anything you say." Which with a known liar is kind of the default, but that is kind of a psychosis, meaning that it blurs your cultural, historical, temporal understanding of things when you're given nothing but garbage, how do you then come up with a response that's accurate? Garbage in, garbage out.[16]

Since actual psychosis would mean that he was suffering from the kind of clinical delusions that I defined back in Chapter 1, Jones's characterization of himself as having had a "form of" or "kind of" psychosis instead suggests the self-awareness that he might have fallen victim to the other kind of false beliefs discussed throughout this book—what I've referred to as "delusion-like beliefs"—that have little if anything to do with mental illness. And yet, without interviewing Jones face to face as I did with the Drummonds, it's impossible for me to know the extent to which he might have really been led astray by mistrust and misinformation as he seems to claim.

A flattering *Rolling Stones* exposé on Jones reported that Jones's "most enduring influence" was reading the 1971 bestseller *None Dare Call It Conspiracy* as a teenager, which he called a "primer to the New World Order."[17] While this suggests that he might have started as a passive consumer of misinformation at an early age, he has since ascended the food chain to become an apex predator and opportunistic misinformer within the disinformation industrial complex, creating novel content and seemingly spewing unfounded conspiracy theories off the top of his head while stoking mistrust of institutions of epistemic authority in the process. Whether or not he actually believes what he says on his show, his claim about having "almost like a form of psychosis," with false beliefs rooted in mistrust and misinformation as opposed to mental illness, is a reasonable characterization of the underlying psychology of his listeners as well as an account of his financial success. Like he said, garbage in, garbage out.

At its core, the disinformation industrial complex that Jones inhabits is built on a model of profit at the expense of truth and a kind of demagoguery that claims, "no one deserves to be trusted, but trust me" and "don't buy what anyone else tells you, but do buy what I'm selling." Jones has said that he uses the substantial profits from the sale of nutritional supplements and survival gear for the coming apocalypse that he offers on his website to fund his revolution,[12] but the reverse seems more accurate—it's really the disinformation he spreads that generates the necessary advertisement and consumer appetite for his snake oil–selling empire.[13,18] Whether we're talking about Jones, Newsmax, OANN, or someone with a YouTube channel claiming the Earth is flat, opportunistic misinformers have capitalized on a market for selling falsehoods to an audience that's repeatedly told that they're being lied to by everybody else.

Just Asking Questions

A few years ago, when I was keeping an eye out for new employment opportunities, one particularly intriguing job posting caught my attention. The job title was advertised as a "research scientist/project manager," and the work description specified "ensuring that brand partners, vendors, and their respective products meet the quality and business standards" set forth by the company. The potential employer was Goop, Inc., the "wellness and lifestyle" empire of actress-turned-CEO Gwyneth Paltrow.

The posting appeared not long after Goop had been in the news for settling a $125,000 lawsuit over false advertisements for some of its products, most notably its infamous $66 jade egg. The egg, which was intended to be inserted into one's vagina, was marketed as being able to correct hormonal imbalances, boost orgasms, prevent uterine prolapse, and regulate menstrual cycles, but the lawsuit contended that none of those claims was "supported by competent and reliable scientific evidence."[19] In addition to the monetary settlement, anyone who'd purchased a jade egg from Goop during an 8-month stretch in 2017 was offered a full refund.

Given that recent development, I considered for a brief moment whether Goop might be genuinely interested in doing legitimate research to test its products—were they looking for help to design and implement a randomized, controlled trial of the jade egg? I was skeptical, but couldn't help but wonder what it might be like to apply for the job, if only to understand what they might actually mean by the term "research scientist" or to have something to write about. Needless to say, I didn't apply, but within the next year, my skepticism was confirmed by the release of a Netflix series entitled, *The Goop Lab*, around the same time that Goop was back in the news for marketing a $75 candle called, "This Smells Like My Vagina." The six-episode series covered a wide range of topics ranging from female sexuality and bona fide research on the use of psychedelics in psychiatry to practices without any real scientific evidence to support them such as "energy healing" and talking to the dead. With featured "experts" including a chiropractor, a former ER doctor who now practices "functional medicine" (a type of "alternative medicine" with no requirement for a medical doctor degree to become a practitioner), and a PhD in pharmacology and toxicology whose research on psychics has been published in *Journal of Parapsychology*, the show put all the trappings of pseudoscience (a topic I'll return to in Chapter 7) on full display, with New Age hokum dressed up in the guise of legitimate science. However, perhaps having learned from the jade egg lawsuit, each episode began with a disclaimer stating that it was "designed to entertain and inform—not provide medical advice." A review in *The New Yorker* described the show as "either the apex or nadir of infotainment," while another in *Rolling Stone* dubbed Paltrow "the most effective troll of 2020."[20]

As a "wellness brand" launched in 2008 from modest beginnings, Goop has undoubtedly mastered a formula for financial success. In 2016, Paltrow claimed to have raised $10 million from venture capitalists, with revenues reportedly tripling from the previous year.[21] According to *CEO Magazine*, Goop was valued at $250 million as of 2021.[22] Goop's sales profits come from an extensive product line that includes not only jade eggs and curiously scented candles, but perfume, makeup, clothing, jewelry, furniture, bottled water, vibrators, and nutritional supplements. The supplements, containing a mélange of herbs and vitamins, aren't that much different from Alex Jones's except that instead of bearing names like Super Male Virility, they're marketed to women with brand names like Madame Ovary, The Mother Load, and Why Am I So Effing Tired?

Beyond their costly nutritional supplements, Paltrow's Goop and Jones's InfoWars share something else more fundamental to their business plans—they both profit from selling misinformation to a mistrusting consumer base. While InfoWars' audience largely directs its mistrust at government leaders, Goop's wellness-centric audience is one that's disenchanted with Western healthcare's focus on disease and looking for something more, but, in both cases, their audiences are sizeable and growing. As with mistrust in government, Pew polls have found that Americans' trust in the institution of medicine has been in slow decline since the 1970s, with only 37% of US adults agreeing that they have "a great deal of confidence" in it as of 2018.[23] Paltrow, who is said to have become interested in wellness herself after her father was diagnosed with cancer,[21] has tapped into a sizeable market and managed to attract a mostly female consumer base able to pay considerable sums for products that purportedly promote health without any scientific evidence to prove that they actually do. And since mistrust in science is central to Goop's New Age appeal, lack of empirical evidence presents little obstacle to its success.

It could be justifiably argued that the substantial cost of Goop's product line is a small price to pay—to those who can afford it—for what it's really selling, which is nothing less than a healthy and much-deserved dose of female empowerment that's been a long time coming, along with an opportunity for a little psychological self-care in today's harried world. And if that's indeed the only effect of using jade eggs, scented candles, and nutritional supplements, then it hardly matters to those paying for them if such products are merely expensive placebos. In response to both criticism and ridicule for promoting jade eggs and pseudoscience, Goop defended itself by implying that very argument.

We always welcome conversation. That's at the core of what we're trying to do. What we don't welcome is the idea that questions are not okay. Being dismissive—of discourse, of questions from patients, of practices that women might find empowering or healing, of daring to poke at a long-held belief—seems like the most dangerous practice of all. Where would we be if we all still believed in female hysteria instead of orgasm equality? That smoking didn't cause lung cancer? If every nutritionist today saw the original food pyramid as gospel?[24]

And yet, as Amanda Mull—who product-tested $1,279 worth of items purchased at the Goop store in Manhattan including an $80 bottle of water with a quartz crystal in it and wrote about it in *The Atlantic*—"just asking questions" is the standard defense of conspiracy theorists.[25] Indeed, "just asking questions" and the related phrase "do your own research" are declarations of mistrust that implies that the status quo and institutions of epistemic authority are untrustworthy. In Goop's case, "asking questions" suggests that the medical establishment, medical experts, and healthcare professionals don't deserve our trust and that Goop and its own "medical experts" and "research scientists" have something better to offer. That's the same disinformation industrial complex strategy that Jones employs.

To be clear, "asking questions" and "doing your own research" can be healthy manifestations of appropriate skepticism in a world where information and misinformation sit side by side. But when mistrust is driving confirmation bias and motivated reasoning, it's a set-up for being conned by apex predators peddling falsehoods from the top of the disinformation food chain. As Goop contends, there's little doubt that the institution of medicine has been wrong in the past and, even worse, that it has a long history of short-changing women in healthcare. But that doesn't mean that one's health is better off with a jade egg and an $80 bottle of water.

The Anti-Vaccine Disinformation Dozen

In 2021, Goop faced another class action lawsuit after a few consumers reported that the "This Smells Like My Vagina" candle had exploded on them. Explosions and drained wallets aside however, the greatest danger of the kinds of products and practices that Goop sells and promotes—like nutritional supplements, coffee enemas, "vaginal steaming," and "energy healing"—lies in the possibility that some might choose such alternatives in place of medical interventions that are held to a scientific standard of proven effectiveness and stated side effects. In response to Goop's Netflix series, Dr. Nikki Stamp, an Australian surgeon, penned a *Washington Post* op-ed that called Goop a "platform of misinformation, privilege, and anti-science rhetoric" and noted that "what wellness sells is by no means harmless."[26] She cited a 2018 *JAMA Oncology* study that found that among a cohort of nearly 2 million patients with treatable cancer, the use of "alternative medicine" was associated with both a greater likelihood of refusing conventional treatments as well as a two-fold greater risk of death.[27]

There's little evidence that Goop or Paltrow herself advises anyone to turn away from conventional medicine when they're ill. However, the same can't be said for

Dr. Kelly Brogan, a physician who was promoted as a "trusted expert" at a 2018 "Goop health summit," a yearly event that routinely sells out of its top-tier $1,500-a-pop tickets. Brogan spent her undergraduate years studying brain and cognitive science at MIT and went on to complete medical school and a psychiatry residency just like I did, but our paths diverged sharply after that. In 2009, she opened a boutique private psychiatric practice in Manhattan, but soon came to reject the idea that psychiatric disorders are appropriately treated with medications, painting mainstream psychiatry and the pharmaceutical industry as bogeymen. Relabeling herself as a "holistic psychiatrist," she touted replacing psychopharmacotherapy with coffee enemas, Kundalini yoga, a whole-foods diet, and nutritional supplements offered within a subscription-based 44-day online bootcamp that she called "Vital Mind Reset."[28] She even published a randomized-controlled trial demonstrating a significantly greater reduction in depressive symptoms with her "multimodal, online, community based lifestyle intervention" in people "with a history of major depressive disorder" compared to those with no treatment in *Cureus*, an open-access journal that has been described as predatory.[29] While this career path and her achievements to date might sound laudable enough, aligned as they are with the same kind of healthy lifestyle-meets-wellness philosophy promoted by Goop, Brogan began to break more sharply with her medical training to embrace more harmful rhetoric as she transformed her clinical practice in kind while gaining notoriety in the process. In 2014, she infamously dismissed the notion that HIV causes AIDS and suggested that antiviral medications kill more people than the disease.[30] By 2020, during the first months of COVID-19, she was claiming that "there is potentially no such thing as the coronavirus" and went so far as to state that she doesn't believe in "germ-based contagion," seemingly refuting the fundamental fact that germs cause infectious disease.[31]

When I first heard about Brogan a few years ago, I found myself wondering how someone with essentially the same 12 years of higher education that I had could come to believe that germs don't cause disease. Brogan's own telling of her story, along with an exposé by Matt Remski published in *Medium*, offer some clues. Like Paltrow's embrace of wellness after her father's cancer, Brogan's seismic shift away from her medical training started with a personal experience of illness. After being diagnosed with a condition called Hashimoto's thyroiditis while pregnant with her first child, she was treated with medications but decided that she "didn't want to take a prescription for the rest of her life, and could no longer stomach the hypocrisy" of prescribing them to others.[32] While I don't presume to know any additional details of Brogan's medical history, it should be noted that postpartum Hashimoto's thyroiditis is usually, if not always, a self-limiting condition that resolves after pregnancy. The standard treatment involves temporary hormone replacement therapy that restores the body's deficiency in thyroid hormone with naturally occurring thyroid hormone supplements. Nonetheless, Brogan has said that she sought treatment from a naturopath who offered diet changes and nutritional supplements in place of conventional medical treatment and claimed to be cured by the process, which inspired her to adopt the same approach in her own psychiatric practice. Meanwhile, Brogan married Sayer

Ji, a longtime promotor of "alternative medicine" and anti-vaccine disinformation, and, according to Remski, merged business interests as a "wellness power couple" who are "at the forefront of the burgeoning political-religious movement dubbed 'conspirituality,' the strange lovechild of alt-right conspiracists and New Age wellness influences."[28] In a video promoting Vital Mind Reset on her website, Brogan advises would-be clients to "reclaim the power we have given to authorities outside of ourselves that keep us helpless and dependent and in victimhood." Meanwhile, Remski noted that both she and her husband promote the idea of "emancipation" and "sovereignty" in a way that has eerie echoes of the sovereign citizen movement.

A simpler if more cynical explanation for Brogan turning her back on what she learned in medical school is that she and her husband have been trying to capitalize on the same recipe for celebrity and financial success as Jones and Paltrow by peddling mistrust and disinformation. In 2018, according to a client who spoke with Remski, Brogan was charging $4,187 for an initial 3-hour consultation with subsequent 45-minute sessions offered at a rate of $570 an hour in 2018. But she then closed her clinical practice the following year, let her board certification in psychiatry lapse, and now operates "Kelly Brogan MD" as a brand like Goop, with a newsletter, tireless social medial presence including her own YouTube channel, articles in the popular press, two published books including a *New York Times* bestseller, numerous appearances on podcasts like *The Joe Rogan Experience*, and a subscription-based online membership in a "healing community" that she calls the "Vital Life Project" along with "Vital Mind Reset," her more intensive and expensive "step-by-step nutrition and lifestyle protocol."

The merits of "holistic psychiatry" and her use of her "MD" title to legitimize a nonmedical practice aside, Brogan has used her growing notoriety to spread vaccine misinformation. Or perhaps that's better framed in reverse—as illustrated by Jones and Paltrow, spreading controversial misinformation turns heads and creates an audience for what those in the disinformation complex are really selling. In any case, the nonprofit Center for Countering Digital Hate (CCDH) listed Brogan and her husband as 8th and 9th on a list of the "disinformation dozen" most responsible for spreading vaccine-related disinformation on social media.[33] During a 2-month stretch in early 2021, the CCDH found that some 700,000 posts with anti-vaccine content on Facebook could be traced back to those 12 social media influencers. Across both Facebook and Twitter, they were responsible for spreading up to 65% of all anti-vaccine content.

While Brogan may be a rising star within the disinformation industrial complex churning out falsehoods about vaccines, she still has a long way to go to match the efforts and success of those above her on the CCDH's disinformation dozen list. For example, Ty and Charlene Bollinger occupy the number three spot based on their work as "anti-vax entrepreneurs" who manufacture and sell disinformation about cancer, COVID-19, and vaccines, including the claim that Bill Gates was planning to use vaccines to implant microchips in people for the purposes of tracking. According to the Associated Press, the Bollingers' company that markets and distributes their disinformation, TTAC Publishing LLC, has netted $25 million in customer transactions since 2014.[34]

But that's still nothing compared to Dr. Joseph Mercola, the top spreader of vaccine disinformation according to the CCDH. Mercola is an osteopathic physician who once practiced within the traditional medical model but changed his tune in favor of a "natural medicine" approach that included denying that vaccines work while promoting "thermography," nutritional supplements, and all manner of lifestyle products including vegan-waxed dental floss, air and water purifiers, grass-fed meat, organic tampons and cotton underwear, probiotics for pets, and tanning beds that he sells online through his Mercola Market storefront. No matter that the FDA issued warning letters about his unfounded claims that thermography was an effective cancer screening tool and that vitamin supplements can treat COVID-19 or that his tanning beds were discontinued and refunded based on unsubstantiated claims about their ability to "slash your risk of cancer" following a $2.6 million settlement with the Federal Trade Commission.[35] Like Goop's $125,000 fine for jade egg claims, that penalty was a drop in the bucket of Mercola's profits amounting to a self-proclaimed net worth of more than $100 million. With a vast social media following numbering in the millions, a website in operation since 1997 that attracts some 2 million new viewers a month, and numerous media appearances on the likes of *The Dr. Oz Show*, Mercola has been called both a "champion of taking charge of our own health" as well as "the 21st century equivalent of a snake-oil salesman."[36] Echoing Jones, Mercola has claimed that his revenue stream exists to "pay our staff to provide information to educate the public and make a difference,"[37] rather than admitting that the reverse is true—that apex predators at the top of the disinformation food chain exploit the lure of disinformation to hawk their wares, becoming millionaires in the process.

Disinformation Robots

It should be clear by now that the secret to profiting from spreading disinformation is both simple and stereotyped—whether we're talking about Jones, Paltrow, Brogan, or Mercola, their recipe for success is much the same. First, recalling that "fake news" travels farther and faster online than reliable information, apex predators use provocative disinformation to garner media attention and gain a following. Second, they cultivate mistrust in authoritative sources of information and experts under the guise of "just asking questions" while encouraging their audiences to "do their own research" and empowering them to "take charge" and "reclaim power" by listening to what they say. Finally, having won over the trust of people who mistrust experts by convincing them that *they're* the real experts—keeping in mind that the term "con artist" is short for "confidence artist"—they convince their audience that the path to truth and well-being lies in buying the nutritional supplements and other snake oil remedies that they're pushing. In that sense, apex predators are often wolves in sheep's clothing, branding themselves as underdogs fighting the establishment when in fact they're perched comfortably at the top of disinformation food chain.

Although this model of financial success hinges on a charismatic talking head and a cache of vitamins for sale, neither is a requirement for profiting from disinformation. During the 2016 US Presidential election, teenagers and young adults from the small town of Veles in Macedonia stumbled on a get-rich scheme that involved setting up multiple websites and social media personas to tirelessly pump out "sensationalist [and] utterly fake news," mainly appealing to Trump supporters at the time; these attracted so many hits and associated profits from ad revenues that it was described as a "digital gold rush."[38] Following the instruction of Mirko Ceselkoski, a "self-taught, viral marketing specialist," and his online course "Facebook Marketing University," his "students" allegedly netted a "collective $10+ million per month peddling fake news."[39] No talking heads or vitamins needed.

As it turns out, profiting from disinformation doesn't even require the tireless efforts of a living, breathing person at all. Within the social media landscape, "bots"—automated computer programs that churn out posts without a real person behind them—also mobilize disinformation to generate profits. In 2018, George Mason University Professor David Brontiakowski and his colleagues analyzed nearly 2 million Tweets over a 3-year period between 2014 to 2017 and found that Twitter accounts judged to be bots were significantly more likely to post anti-vaccine content than real accounts.[40] This was especially true for bots classified as "content polluters" that used anti-vaccine messaging as "clickbait" to disseminate malware and other unsolicited commercial content for profit. Subsequent research has found that while the number of overall Twitter users who see or share bot-driven vaccine content may be quite small, it's disproportionately greater among those "embedded in communities where exposure to vaccine-critical content is common."[41] Meanwhile, more recent research has shown that Twitter users can't reliably distinguish between tweets written by human beings and those generated by the artificial intelligence (AI) text generator GPT-3 (the large language model used by the AI chatbot ChatGPT) across a variety of topics including vaccine safety, evolution, COVID-19, and climate change.[42] Furthermore, it was more difficult for users to correctly identify tweets as false when they were generated by the text generator compared to when they were written by real users.

Taken together, these findings suggest that, going forward, we can expect to encounter much more disinformation that's mass-produced by machines, whether in the form of tweets by bots, "fake news" articles written by AI, or the kind of "deepfake" videos that I briefly mentioned back in Chapter 3. Such computerized disinformation generators can be likened to an army of pawns within a larger game of chess being played by the apex predators within the disinformation industrial complex. The question that we should all keep asking ourselves is just who might be controlling those pawns.

statement's source increases perceptions of truth, as we might expect, the truth effect persists even when sources are thought to be unreliable and especially when the source of a statement is unclear.[45] In other words, while we typically evaluate a statement's truth based on the trustworthiness of the source, as I noted earlier, repeated exposure to both information and misinformation can increase the impression that it's true regardless of the source's perceived credibility. The illusory truth effect tends to be strongest when statements are related to a subject about which we believe ourselves to be knowledgeable[46] and when statements are ambiguous such that they aren't obviously true or false at first glance.[45] It can even occur with statements that are framed as questions, like "Is President Obama a Muslim?," due to a related phenomenon called the "innuendo effect."[47] While I spent much of Chapter 4 joining Ted Koppel in taking news media to task for circulating misinformation, these aspects of the illusory truth effect mean that even objective news reporters face a difficult quandary when covering politicians who make untruthful statements. While such statements demand legitimate news coverage, printing them in newspaper headlines and repeating them in soundbites across the media landscape amounts to free advertising that inadvertently contributes to belief in disinformation. Given what's known about the illusory truth effect, that's likely to occur even if news reports offer a disclaimer that the statement is untrue.

Another feature of the illusory truth effect that's particularly relevant to disinformation is that it can occur despite prior knowledge that a statement is false[48] as well as in the setting of real-life "fake news" headlines that are "entirely fabricated ... stories that, given some reflection, people probably know are untrue."[49] Research by Gordon Pennycook, whose work on belief in fake news I described in the previous chapter, has determined that the illusory truth effect can even occur with prior and repeated exposure to fake news headlines that run contrary to one's political party affiliation. For example, exposure to a fake news headline like "Obama Was Going to Castro's Funeral—Until Trump Told Him This" has been shown to increase perceptions of its truth not only for Republicans, but for Democrats as well.[49] For good or bad, the illusory truth effect therefore seems to be able to—no pun intended—trump motivated reasoning.

As Goebbels foreshadowed, exploiting the illusory truth effect has become a common and powerfully effective tactic of political propaganda today so that it deserves much greater public awareness. As consumers of information, we should all be cognizant of the fact that any process that increases our familiarity with disinformation—whether through repeated exposure or otherwise—can increase our perception that falsehoods are true. This effect can occur despite being aware that the source of a statement is unreliable, despite previously knowing that the information is false, and despite it contradicting our own political group's "party line." In Chapter 4, I noted that there may be no better way to get people's attention than misinformation. The illusory truth effect suggests that there's no better way to get people to believe misinformation and to disbelieve the truth than to simply keep repeating it over and over again.

Political Propaganda and the Illusory Truth Effect

Earlier I noted that the typical motives for apex predators to spread disinformation are to gain either financial profit or political power. Acknowledging that profit and power are rarely mutually exclusive goals, I'll focus on the use of disinformation for political motives for the remainder of this chapter. Misrepresenting the truth for political motives is so common that we have a specific name for this kind of disinformation—"propaganda." Although the term has been around since at least the 1700s, it has come, over the past century, to refer more specifically to disinformation intended to advance a political cause, coincident with World Wars I and II.

In *Mein Kampf,* Adolf Hitler infamously coined the term "*große Lüge*" *or* "big lie" to describe a falsehood so "colossal" and outlandish that people would believe it on the grounds that no one would imagine that anyone would ever lie so boldly. Although Hitler wasn't referring to himself, Joseph Goebbels, the Nazi Party's Minister of Propaganda, has been credited with a similar and now familiar quotation, "repeat a lie often enough and people will eventually come to believe it." The Nazi Party's big lie—also known as the "stab in the back myth"—was the baseless claim that Germany hadn't suffered defeat in World War I per se, but had instead been betrayed internally by domestic traitors in the form of socialists and Jews. This specific piece of propaganda, coupled with other claims about the greatness of Hitler and the Nazi Party, are thought to have been crucial in creating enough public support to seize power, embark on a plan to exterminate Jews and other minority groups, and invade Poland to start World War II.

Goebbels's claim that people are likely to believe lies that are repeated often enough has been validated by decades of research on what psychology calls the "illusory truth effect." First described in a 1977 study by Temple University psychologist Lynn Hasher and her colleagues, the illusory truth effect occurs when repeating a statement increases the belief that it's true even when the statement is actually false.[43] Subsequent research has expanded what we now know about the effect. First of all, it doesn't only occur through repetition, but can also happen through any process that increases familiarity with a statement or the ease by which it's processed by the brain—what psychologists refer to in this context as "fluency." For example, the perceived truth of written statements can be increased by presenting them in bold, high-contrast fonts or when aphorisms are expressed as a rhyme.[44] Indeed, bold product claims and catchy slogans or songs, regardless of their veracity, have always been a standard and effective component of commercial advertising, where the terms "puffing" and "puffery" refer to baseless claims about a product that, despite leaving a company liable to false advertising litigation, often remain profitable in the long run.

According to a 2010 meta-analytic review of the more generic "truth effect" that applies to both true and false statements, while the perceived credibility of a

The Firehose of Falsehood

In the classic novel *1984*, George Orwell portrayed a fictitious dystopia inspired by Stalin's Soviet Union in which a totalitarian political party oppresses the public through "doublethink" propaganda epitomized in the oxymoronic slogan, "War is Peace, Freedom is Slavery, Ignorance is Strength."[50] "Doublethink," Orwell wrote, consists of "the habit of impudently claiming that black is white, in contradiction to plain facts … it means also the ability to *believe* that black is white, and more to *know* that black is white, and to forget that one has ever believed the contrary." In the storyline of *1984*, this is achieved through a constant contradiction of facts and revision of history to the point where people are left with little choice but to resign themselves to accept party propaganda: "The party told you to reject the evidence of your eyes and ears. It was their final, most essential command." Recalling the misinformation effect—the ability to modify memories through cueing—that I discussed in Chapter 2, this kind of psychological "brainwashing" isn't so far-fetched.

The concept of "doublethink," along with Orwell's portrayal of the state-controlled language "Newspeak" in *1984* gave rise to the now familiar amalgamated term "doublespeak" to describe a way of talking—often in politics—that deliberately obscures and misrepresents the truth. Deceptive doublespeak has been a well-characterized propaganda tool of the Soviet Union—the official newspaper of the Soviet Communist Party for nearly a century was unironically called *Pravda* or "Truth"—that continues within modern Russia today. In 2016, RAND researchers Christopher Paul and Miriam Matthews described the Russian model of propaganda as a "firehose of falsehood" due to a core strategy that hinges upon the "shameless willingness to disseminate partial truths or outright fictions" in a way that's "rapid, continuous, and repetitive."[51] They cited the illusory truth effect in explaining how this strategy has been so surprisingly effective through a barrage of disinformation and a "hit first" mentality that cultivates familiarity and a false sense of credibility.

At the core of Russia's political propaganda machinery is the Internet Research Agency (IRA), a Kremlin-backed enterprise that harnesses disinformation along with the power of social media to manipulate public opinion across the world according to Russian political and corporate interests. The IRA's strategy centers around the use of a veritable army of internet "trolls" or deceptive social media accounts with fake personas who post all manner of content with the intention of fomenting discord. According to interviews with those who have worked within such "Russian web brigades" and "troll farms," the IRA has different "desks" focusing on diverse agendas both within Russia and abroad, including a "Department of Provocations" whose goal is to "set Americans against their own government: to provoke unrest and discontent."[52] It accomplishes this not only by posting disinformation in the form of false news stories and hoaxes, but also by playing both sides of controversial topics and political debates with "right trolls" and "left trolls"—that is, fake personas on either side of the political fence.[53] Over the past several years, the favored content of the IRA's trolling has been the US Presidential elections of 2016 and 2020; police brutality and the Black Lives Matter movement; lesbian, gay, bisexual, transgender, and queer (LGBTQ) rights; issues related to immigration; and vaccines—the very same divisive, hot button issues that by design have been fueling political polarization in the United States and tearing apart the fabric of America in the process.

Just how much the IRA, or the Macedonian troll farms mentioned earlier, deserve credit for exacerbating political polarization or actually swaying the course of US elections to date remains unclear. Several research studies have found that online engagement with content posted by the likes of Russian trolls may be quite limited and that most who do engage with it are already highly polarized.[54] However, a report by the Computational Propaganda Research Project found that, between 2013 and 2018, tens of millions of US users were exposed to IRA-generated content across Facebook, Twitter, and Instagram, with some 30 million sharing it online in the three years between 2015 and 2017.[55] In 2017 alone, the IRA was responsible for nearly 60,000 Twitter posts per month and 541 paid ads per month on Facebook. And while the IRA represents one of the largest efforts to industrialize disinformation, such efforts are hardly unique to Russia—as of 2019, some 70 countries operate similar disinformation campaigns, including the United States.[56]

Because disinformation is passed down from trolls to real people along the disinformation food chain, there's often little hope of readily identifying its origins. Like money laundering, such "information laundering" means that disinformation often ends up integrated with and indistinguishable from reliable information.[57] In much the same way, information laundering has paved the way to political "astroturfing" campaigns that give the illusion of grassroots public support, such as those that drum up enthusiasm for social causes and even manage to get people to show up at rallies and marches while concealing the vested interests of those funding the campaigns.[58] As consumers of online information, we would therefore do well to always keep in mind that what we encounter on the internet could very well be generated by shadowy figures who are trying to incite our passions, increase the righteousness we feel about our beliefs, erode trust in traditional institutions of epistemic authority, and egg us on to fight with our neighbors.

Alternative Facts in a Post-Truth World

Over the past few decades, political disinformation strategies have shifted away from merely spreading falsehoods to breeding mistrust in facts and experts, thereby eroding the very concept of truth. Some have therefore claimed that, as a result, we have come to inhabit a "post-truth" world.

Peter Pomerantsev, a Soviet-born journalist and author of *This Is Not Propaganda: Adventures in the War Against Reality*, first observed this shift in post-Soviet Russia during the decade he spent working there from 2001 to 2010.

> It's not about proving something, it's about casting doubt.... this new propaganda is different. Putin isn't selling a wonderful communist future. He's saying, we live in a dark world, the truth is unknowable, the truth is always subjective, you never know what it is, and you, the little guy, will never be able to make sense of it all—so you need to follow a strong leader.[59]

Now living in London, Pomerantsev has witnessed the same shift in propaganda spread around the world, including here in the United States.

> Instead of trying to argue in a rational way, politicians become these great performance artists, trying to be outrageous, reveling in the fact that they don't care about the facts....
> The same kind of politics I saw in Russia years ago is the same kind of politics I'm seeing now in the UK and Brazil and the Philippines and the US. And the internet and digital media technologies have been the essential tools behind it.[59]

Back in 2011, before he formally tossed his hat into the political ring, Donald Trump established himself as the spokesperson of the "birther" conspiracy theory movement, going on the talk show circuit to question whether President Barack Obama was born in the United States. A national survey from that year revealed that 24% believed that Obama wasn't born in the United States, with an additional 24% neither agreeing or disagreeing.[60] In retrospect, this wasn't only a powerful illustration of the illusory truth effect in action, but a harbinger of things to come. After Trump won the 2016 US Presidential election by a sizeable margin of the electoral vote, he tweeted that he'd also won the popular vote, a statement that flew in the face of objective reality—that he'd lost it by more than 2.8 million votes. In response to criticism for that false claim, Trump campaign surrogate and self-described "journalist and patriot" Scottie Nell Hughes defended it thusly:

> ...one thing that has been interesting this entire campaign season to watch, is that people that say facts are facts—they're not really facts. Everybody has a way—it's kind of like looking at ratings, or looking at a glass of half-full water. Everybody has a way of interpreting them to be the truth, or not truth. There's no such thing, unfortunately, anymore as facts.
> And so Mr. Trump's tweet, amongst a certain crowd—a large part of the population—are truth. When he says that millions of people illegally voted, he has some facts—amongst him and his supporters, and people believe they have facts to back that up. Those that do not like Mr. Trump, they say that those are lies and that there are no facts to back it up.[61]

Soon thereafter, President Trump would go on to boast about the size of the crowd that attended his inauguration noting that it "looked like a million-and-a-half people," with his new Press Secretary Sean Spicer claiming that it was "the largest audience ever to witness an inauguration, period, both in person and around the globe."[62] When that hyperbole was refuted by Chuck Todd on an episode of NBC's *Meet the Press*, Senior Counselor to the President Kellyanne Conway claimed that Spicer's claims represented "alternative facts" to which Todd replied, "alternative facts are not facts ... they're falsehoods."[63]

Throughout his four years of office, President Trump repeatedly denounced the "fake news media" as the "enemy of the people." Meanwhile, while maintaining a running count, *The Washington Post* tallied a total of 30,573 falsehoods and misleading claims made by Trump over the course of his presidency.[64] For example, on more than 150 occasions, he repeated the claim that his administration passed the "Veterans Choice" program that allowed veterans to access care outside of the Veterans Healthcare Administration (VHA), even though in reality the Veteran's Access to Care Through Choice, Accountability, and Transparency Act was passed under the Obama administration in 2014.[65] Then, as many suspected he would in the wake of being defeated by Joe Biden in the 2020 election, Trump went out with a bang, claiming what would come to be described as his own "big lie"—that the election was rigged, that votes for Biden were fraudulent, and that he hadn't actually lost at all. No matter that this claim was wholly unsupported by evidence, with some 60 lawsuits purporting election fraud either abandoned or overturned. On January 6, 2021, after Trump told his supporters, "we will never give up, we will never concede ... we will not take it anymore ... we will stop the steal ... we don't have free and fair elections,"[66] an angry mob that bought the big lie stormed the US Capitol in an act that was part protest, part riot, and part insurrection. Six months later, as much as a third of the American public and two-thirds of those identifying as Republicans still believed that the election had been stolen.[67] As of 2023, those numbers remained largely unchanged, despite increasing acknowledgment that there's no evidence to support such claims.[68]

It's hard to imagine that Vladimir Putin, the apex predator of political disinformation, would be feeling anything but the smug satisfaction of a mission accomplished upon hearing an American president declare that our elections aren't free and fair and witnessing his plan to "set Americans against their own government" come to fruition. Perhaps Putin also enjoys a kind of parental pride in the fact that the Trump administration seemed to emulate Russia's firehose of falsehood, calling objectively measurable facts into question time and again and eroding the very concept of truth in the process. Indeed, at the start of Trump's four years as president, Conway's mention of "alternative facts" suddenly catapulted Orwell's *1984*—a novel written in 1949—to the top of Amazon's bestseller list. Later, when she was pressed to account for what she meant by that dubious phrase, Conway explained "alternative facts" as "additional facts and alternative information" while noting,

> Two plus two is four. Three plus one is four. Partly cloudy, partly sunny. Glass half full, glass half empty. Those are alternative facts.[69]

That kind of disturbingly Orwellian doublespeak, along with the idea that facts and truth are subjective and might not even exist at all come straight out of the Russian propaganda playbook. Just so, it's impossible to ignore the fact that after he lost the 2020 election, Trump focused his energies on the launch of Truth Social, his own social media platform with a name disturbingly reminiscent of the Soviet Union's *Pravda*.

Of course, those who believe the election was stolen also believe that it's the Democrats and liberals that are spreading lies, just as those who are fans of Paltrow's Goop and Mercola's "natural medicine" are wont to claim that it's the medical establishment—in bed with Big Pharma—that is the apex predator at the top of the disinformation food chain. After all, they argue, all politicians and presidents lie, regardless of party affiliation. In 1988, President George H. W. Bush told us, "read my lips: no new taxes." To justify war with Iraq, President George W. Bush, along with British Prime Minister Tony Blair, claimed that Saddam Hussein had weapons of mass destruction. In 1998, President Bill Clinton insisted, "I did not have sexual relations with that woman." President Obama promised us that he would close the prison at Guantánamo Bay and that his universal healthcare bill would allow everyone to keep their own physicians and not increase healthcare costs. And why pick on someone like Gwyneth Paltrow when it's Big Pharma that's using big lies to promote corporate self-interests? Just look at how Purdue Pharmaceuticals plotted to create the opiate epidemic!

By themselves, such counterclaims are perfectly valid. The few examples of apex predators that I singled out for this chapter were selected for their potential to undermine the integrity of our democratic institutions, incite political violence, and steer people away from proper medical care, but they're admittedly but a few drops in a deep sea of disinformation. At the same time, "whataboutism" that ends with the conclusion that "everyone lies" rather than agreeing that honest sources of information—and facts themselves—do exist demonstrates the very intent of the disinformation industrial complex's post-truth dogma. Hannah Arendt, a political theorist and foremost authority on the makings of modern totalitarian regimes, summed up the manipulative goal of post-truth politics as follows:

> If everybody always lies to you, the consequence is not that you believe the lies, but rather that nobody believes anything any longer.... And a people that no longer can believe anything cannot make up its mind. It is deprived not only of its capacity to act but also of its capacity to think and to judge. And with such a people you can then do what you please.[70]

If we can be convinced that facts are always open to interpretation and endlessly debatable, inconvenient truths can then be dismissed as "fake news." If we can't agree as a society about who's lying and who's telling the truth, lies can be cited as evidence and objective truths are pushed beyond our grasp. And if truth doesn't exist, we're justified in arguing and fighting with one another rather than ever doing what it takes to establish consensus and work together toward a common cause.

* * *

In 2017, University of Western Australia psychologists Stephan Lewandowsky and Ullrich Ecker, together with George Mason University psychologist John Cook published an academic paper that painted the following portrait of a post-truth world:

> Imagine a world that has had enough of experts. That considers knowledge to be "elitist." Imagine a world in which it is not expert knowledge but an opinion market on Twitter that determines whether a newly emergent strain of avian flu is really contagious to humans, or whether greenhouse gas emissions do in fact cause global warming, as 97% of domain experts say they do. In this world, power lies with those most vocal and influential on social media: from celebrities and big corporations to botnet puppeteers who can mobilize millions of tweetbots or sock puppets—that is, fake online personas through which a small group of operatives can create an illusion of a widespread opinion. In this world, experts are derided as untrustworthy or elitist whenever their reported facts threaten the rule of the well-financed or the prejudices of the uninformed.[71]

A flea market of opinion. Trolls, bots, and astroturfing. A disinformation industrial complex. If that sounds familiar by now, it's because their vision made quite an impact on me at the time and, as I re-read it now, I see that it clearly acted as a kind of wake-up call and a seed crystal for this book. But while Lewandowsky, Ullrich, and Cook noted that the term "post-truth" was virtually unheard of five years prior and that we might not have been living in such a world yet back when they wrote their paper, I think it's safe to say that if we weren't yet there in 2017, we certainly are now. That inescapable conclusion mirrors what RAND researchers Jennifer Kavanagh and Michael Rich found in their analysis that I discussed in Chapter 4—that the extent to which there's now a fundamental disagreement over facts is unprecedented in nearly 150 years of American history.

An Ipsos poll conducted in January 2021 found that 86% of respondents were either "somewhat" or "very" concerned about the spread of false information, 76% doubted the accuracy of information encountered on social media, and 77% were worried about the possibility of political violence in the coming four years.[72] Perhaps surprisingly, those rates were spread evenly between Democrats, Republicans, and Independents. While it's encouraging that we're *all* concerned about misinformation, the fact that we *are* all concerned also highlights the core dilemma—that we don't agree on what's misinformation and what's not.

In a truly post-truth world, the death of facts should mean, as Arendt warned, that we don't believe in anything anymore. But that's not where we are, at least not yet. We all *want* to believe. We all *need* to believe. And so, when we no longer believe in facts, we instead turn to belief in misinformation and lies. As US Naval War College professor and "Never Trump" conservative Tom Nichols argued in his 2017 book *The Death of Expertise: The Campaign Against Knowledge and Why It Matters*, what lies at the heart of our modern inability to agree on facts is the insidious and often deliberate erosion of trust in experts and institutions of epistemic authority.[73] As I've argued throughout this chapter, it's that same mistrust that leaves us vulnerable to manipulation by the apex predators atop the disinformation food chain. Until the day that trust in institutions of epistemic authority is restored, we the global citizens of a post-truth world will remain vulnerable to false beliefs stemming from disinformation and resistant to evidence and fact-checking corrections so that endlessly bickering with each other over facts and alternative facts will continue to split us further apart.

Although we find ourselves living in a post-truth world today, this state of affairs should be recognized as more of a deliberate propaganda strategy than an unavoidable reality. We don't have to allow ourselves to be its victims. If there's room for hope, it lies in understanding just what's going on and who the players are in the disinformation food chain. Although this chapter has focused on disinformation, Lewandowsky, Ullrich, and Cook—along with Colleen Seifert, whom I quoted earlier—make clear that our post-truth world is rooted in both "misinformation in the head" and "misinformation in the world." If we are to fully understand belief in misinformation, we must appreciate how those two roots are intertwined. In doing so, we would benefit from acknowledging the distributed responsibility for disinformation that harms, as I mentioned at the beginning of the chapter. The 2022 jury decision that ordered Alex Jones to pay $965 million in damages to the families of the children who died at Sandy Hook—assuming he ever pays—is a step in the right direction. So is the $787 million settlement that Fox News agreed to pay to Dominion Voting Systems in 2023, over the news network's false claims that Dominion's voting machines had been rigged to steal the 2020 Presidential election.

And yet, if we are to rescue ourselves from the perils of post-truth politics, we must also embrace a more compassionate view of the victims of disinformation who won't let go of false beliefs that fly in the face of facts. Those who fall prey to disinformation aren't our enemies—our common enemies are those within the disinformation industrial complex that exploit our cognitive vulnerabilities to sell us on lies that put us at each other's throats.

6
Conspiracy Theories Gone Wild

> Conspiracy theories are a genre of science fiction in which most organizations are secretly run by competent people pursuing definite goals.
> —Byrne Hobart

Flat Earthers

During the 2017 NBA All-Star Weekend in New Orleans, the Cleveland Cavaliers were the defending champions, having prevailed over the Golden State Warriors in a Game 7 thriller of the Finals during the previous summer. For the All-Star Game, fans voted in the Cavs' "big three" franchise players LeBron James, Kyrie Irving, and Kevin Love to play for the Eastern Conference All-Stars Team. With both East and West abandoning defense even more than usual, the final score of 192 to 182 broke the record for the highest-ever score in an NBA All-Star Game, with the Cleveland players ending up on the losing side of history.

But nothing that went down that weekend related to basketball could top the viral news from just a few days before. During an interview between Irving and his teammates on the *Road Trippin'* podcast, the conversation came around to space aliens and the question of whether the Earth is round or flat. Irving offered this answer:

> This is not even a conspiracy.... The Earth is flat.... All these things that particular groups, I won't even pinpoint one group, that they almost offer up this education. The fact that in our lifetimes that there are so many holes and so many pockets in our history....
>
> Is the Earth flat or round? I think you need to do research on it.... It's right in front of our faces, I'm telling you, it's right in front of our faces. They lie to us....
>
> What I've been taught is that the Earth is round ... but I mean, if you really think about it from a landscape of the way we travel, the way we move and the fact that, can you really think of us rotating around the sun and all planets aligned, rotating in specific dates, being perpendicular with what's going on with these planets....
>
> There is no concrete information except for the information that they're giving us. They're particularly putting you in the direction of what to believe and what not to believe. The truth is right there, you just got to go searching for it. I've been searching for it for a while.

What it really came down to for me was, everything that was particularly thrown in front of me, I had to just be like, "Okay look, this is all a facade." Like, this is all something that they ultimately want me to believe in.[1]

In coming out as a "flat Earther," Irving joined the celebrity ranks of *The View* host Sherri Shepherd, rapper B.o.B., and reality TV personality Tila Tequila along with a broader movement of naysayers that has recurred throughout history. While it's often said that flat Earth beliefs persisted through the Middle Ages until Christopher Columbus set us straight in 1492, a more accurate accounting reveals that enlightened civilizations had accepted the roundness of the Earth more than a millennium before that thanks to the observations and calculations of astronomers and mathematicians like Pythagoras, Euclid, and Ptolemy. And yet, from time to time since then, small pockets of public denialism have emerged to challenge this basic fact as a kind of broader revolt against scientific thinking. In 1956, Samuel Shenton founded the International Flat Earth Research Society, and, despite astronauts producing photographs of a round Earth taken from the moon landing in 1969, the group's official membership peaked to some 3,500 members in the 1990s.[2] With like-minded individuals able to find each other on the internet since then, a number of different "Flat Earth Societies" have managed to gain followings online numbering in the hundreds of thousands. In recent years, flat Earth conventions have been held in the United States, where 16% of the public and as many as 34% of Millennials lack confidence that the Earth is round, as well as internationally in the United Kingdom, Italy, and Brazil.[3]

How is it possible that anyone, much less Irving who spent his freshman year at Duke University, can refuse to believe such a basic science fact? One simple answer is naïve realism that, as I explained in Chapter 1, amounts to the idea that "perception is reality" or that "seeing is believing." Indeed, both B.o.B and Tila Tequila defended their claims by noting that the Earth appears flat on the horizon.[4] But in an era when we now have easy access to photographs and video taken from satellites and the International Space Station that clearly demonstrate the Earth's roundness, there's more to it than that. Despite Irving's insistence to the contrary, if we are to understand flat Earthers, we have to appreciate flat Earth beliefs as a specific kind of claim that seems increasingly common these days—conspiracy theories.

Conspiracy theories reject authoritative accounts of reality in favor of some plot involving a group of people with malevolent intent that's deliberately kept secret from the public.[5] This definition highlights that flat Earth dogma doesn't only include an idiosyncratic belief in the shape of the Earth and a dismissal of the mountain of evidence that refutes it—that, by itself, wouldn't constitute a conspiracy theory—but also the corresponding belief that the National Aeronautics and Space Administration (NASA) is lying to us and that every other country with a space program, along with astronomers and astrophysicists the world over, are also colluding to hide the truth with no discernible rational motive. Debates over faked NASA photos, mathematical proofs, and how it would be possible to circumnavigate the Earth were it not a globe aside, the denial of such a basic scientific fact along with the counter-claim of such an impossibly vast cover-up is what makes flat Earth beliefs so seemingly incredible and therefore intriguing to me as a psychiatrist.

* * *

Over the first half of this book thus far, I've provided an overview of the core building blocks of false belief including naïve realism, overconfidence, confirmation bias, motivated reasoning, mistrust, and exposure to misinformation. In the next few chapters, I'll focus on how those ingredients are relevant to specific types of false belief, starting in this chapter with the phenomenon of conspiracy theories as a timely example of delusion-like beliefs that are surprisingly common; have very little if anything to do with mental illness, intelligence, or education; can be better understood as byproducts of a universal vulnerability to mistrust and misinformation; and have become imminently consequential to the world today.

Before we delve deeper into the psychology of conspiracy theories however, we should first revisit the idea that belief conviction occurs along a continuum, as I discussed in Chapter 1. Since our beliefs are probability judgments that rarely if ever warrant all-or-none conviction, we need to first ask ourselves just how much people like Irving really believe that the Earth is flat. When he was challenged following his initial comments and asked whether he'd seen pictures from space that might lead him to concede that the Earth is round, Irving replied, "I've seen a lot of things that my educational system has said that was real and turned out to be completely fake, so I don't mind going against the grain."[6] But two years later, after high school teachers complained that his comments had set back their students' educations, he apologized that he "didn't realize the effect" his words would have[7] and suggested that he was either being deliberately provocative or didn't know what to believe.

> I don't know. I was never trying to convince anyone that the world is flat. I'm not being an advocate for the world being completely flat. No, I don't know. I really don't. It's fun to think about though. It's fun to have that conversation. It is absolutely fun because people get so agitated and mad. They're like, "Hey man, you can't believe that, man. It's religious, man. It's just science. You can't believe anything else. O.K.?" Cool, well, explain to me. Give me what you've known about the Earth and your research, and I love it. I love talking about it.
>
> … I haven't convinced myself all the way like everything that has been given to us is fake. No. But you also know that a lot of history has been distorted over time. That's something that I'm always aware of.[8]

In a similar way, looking more closely at the poll that found that 16% of Americans and 34% of Millennials lack confidence that the Earth is round, only 2% and 4%, respectively, actually endorsed a firm, enduring belief that the world is flat.[9] The rest were either skeptical or weren't sure what to believe. That 34% of Millennials lack confidence that the Earth is round is still remarkable and concerning in its own right, but it's understandable when we consider that the saying "never trust anyone over 30" has been popular among young people at least as far back as the 1960s. Today, people of all ages find themselves in a post-truth world where political leaders and celebrity icons alike are telling us not to trust anyone, to questions facts at every turn, and to "do your own research" instead of heeding experts in order to find out what's true. No wonder there's so many who are "just asking questions" but have no reliable compass to guide what, or who, to believe.

A Dark Age of Conspiracy Theories

With flat Earth beliefs refuting a basic science fact that's been widely accepted for thousands of years and Obama "birthers," anti-vaxxers, Sandy Hook denialists, and QAnon adherents seemingly popping up all around us, many have claimed that we're living in a "Golden Age of Conspiracy Theories." It's certainly true that belief in conspiracy theories is more common than many of us might suspect. Over the past few decades, surveys have consistently reported that about half the population in the United States and other countries around the world believes in at least one.[10] Remarkably, a 2019 YouGov poll that asked about belief in a number of different specific conspiracy theories found the proportion of conspiracy theory believers to be as high as 72% in Japan; 77% in France; 79% in Great Britain; 80% in the United States, Australia, and Canada; 85% in Spain; and 91% in Mexico.[11] In addition, psychology research tells us that belief in one conspiracy theory predicts belief in others, suggesting that most of us have something of a propensity for conspiratorial thinking or a preference for conspiratorial narratives that psychologists refer to as "conspiracy mentality."[12]

And yet the available evidence falls short of supporting the popular impression that conspiracy theories are any more prolific or more widely believed now than they were in the past. Conspiracy theories—as well as real-life conspiracies—have likely been around since the dawn of civilization. They were certainly present throughout recorded history, whether in Medieval times when they were used to bring about the fall of the Knights Templar, in the 1700s when talk of the Illuminati intent on creating a New World Order first began, or at the turn of the century when the fictitious work *The Protocols of the Elders of Zion* fueled long-standing flames of antisemitism. Over the past 80 years, conspiracy theories circulating about Roswell, the moon landing, the assassination of JFK, the death of Princess Diana, and 9/11 have all been part of mainstream public consciousness.

In their 2014 book *American Conspiracy Theories*, University of Miami political scientists Joseph Uscinski and Joseph Parent reviewed more than 100,000 letters to the editor of *The New York Times* dating back to 1890 to test the claim that conspiracy theory beliefs are more common today than they were over a century ago.[13] Based on a subsample of 635 letters referencing conspiracy theories averaging to about five per year through 2010, they concluded that while there have been ebbs and flows over time—with two sharp spikes in the 1890s when there was a backlash against big business monopolies and in the 1950s during the McCarthy era communist "Red Scare"— conspiracy theories aren't any more prolific today compared to other points in history. A more recent study by Uscinski and his colleagues likewise found no evidence of any net increase in specific conspiracy theory beliefs in the United States from as far back as 1966 to 2020 and no increase in generalized conspiratorial thinking from 2012 to 2021.[14] With these two studies being the only ones of their kind, it's hard to justify the impression that we're living in a Golden Age of Conspiracy Theories. Indeed, although journalists have been declaring the "year of the conspiracy theory" since at least the 1960s, Uscinski derides the claim that "now is the time of conspiracy theories" as the "biggest myth [about conspiracy theories] that keeps getting repeated."[15]

Still, in the absence of carefully and consistently worded surveys that quantify conspiracy theory belief conviction over time, it remains possible that the degree to which people believe conspiracy theories might still be on an upswing. But even if it isn't, a look at polling data from the past few years still provides plenty cause for concern. As I noted in Chapter 4, the COVID-19 pandemic claimed more than 4.5 million lives worldwide by the fall of 2021, including 1 out of every 500 people in the United States. Despite an 11-fold lower rate of death among those vaccinated, about half of Americans were still unvaccinated, confirming what polls about willingness to accept a vaccine once it became available foretold throughout the previous year.[16] It was much the same, if not worse, across Europe. While such "vaccine hesitancy" has been a long-standing challenge with many causes, there's little question that belief in anti-vaccine and COVID-19 conspiracy theories contributes.[17] An *Economist/YouGov* poll from July 2021 found that 40% of the American public believed that the threat of COVID-19 was exaggerated for political reasons.[18] One in five believed that the US government was using the COVID-19 vaccine to microchip the population. Nearly as many believed that vaccines cause autism, while an earlier survey found that 20% also believed the conspiracy theory that "doctors and the government" were aware of this risk but had been hiding it from the public for years.[19]

In 2018, those who voted for the United Kingdom to leave the European Union were substantially more likely to believe in conspiracy theories related to immigration and the integrity of the voting system for the referendum compared to those who voted to stay.[20] Given that "Brexit" passed by only the narrowest of margins, this major geopolitical event probably wouldn't have occurred were beliefs in conspiracy theories not dictating voter behavior to some degree.[21] Here in the United States, following one of the most hotly contested Presidential elections in American history, 40% of the public believed that millions of illegal votes had been counted as of the summer of 2021.[17] Nearly a third of all Americans and two-thirds of Republicans believed the conspiracy theory that the election was stolen by President Biden.[17,22] More than 70% believed that violence due to the election outcome was at least somewhat likely in the future, and 40% of Republicans believed it would be justified and possibly necessary.[23] Fifteen percent believed the QAnon conspiracy theory that the government is controlled by Satan-worshipping pedophiles running a child sex trafficking ring, and 20% believed that a "storm" was coming that would "sweep away elites in power and restore the rightful leaders."[24] And while QAnon may have started here in the United States, it quickly transcended American politics, with its dogma spreading beyond our shores as a part of a worldwide revolt against governmental authority, elites, and experts. Finally, amid all the political chaos, an overwhelming consensus of scientific experts has been warning us about the reality of anthropogenic climate change for the better part of the past decade, but, despite one record-breaking heat wave after another, nearly half of Americans in 2021 didn't believe that climate change is a result of human activity, and as many as 27% believed it to be a hoax (I'll return to the topic of climate change in Chapter 9).[25]

And so, while conspiracy theories might not be any more prolific today, it would be harder to argue that they haven't become more consequential now than they have been over the course of most of our lifetimes. Of course, some conspiracy theory beliefs are harmless—turning back the clock on science education aside, flat Earth conspiracy theories are an otherwise fairly innocuous peculiarity. The same could be said of conspiracy theories related to UFOs, the moon landing, or even the assassination of JFK. But belief in modern conspiracy theories related to vaccines, the integrity of democratic elections, and climate change has a much greater practical relevance to our daily lives and a much greater potential for harm. As I've been suggesting over the past two chapters, belief in disinformation-promoting conspiracy theories about these pivotal issues of our time is already contributing to human lives lost through neglect of proper medical care and failure to take action on attempts to halt or reverse global warming. In that sense, as I claimed in the Preface, we're dying from suicide by false belief. Our collective attraction to conspiracy theories is a big part of the problem.

* * *

Over the past decade, a series of research studies based on survey and questionnaire data—many of them conducted by Northumbria University social psychologist Daniel Jolley—have been exposing the many potential harms of conspiracy theories. Such research has found that conspiracy theory belief is associated with opting out of healthy behaviors like getting vaccinated, both political apathy and extremism, anger, fractured interpersonal relationships, reduced motivations to lower one's carbon footprint, intentions to commit crime, and justifications for violence.[17,19,26]

Meanwhile, outside of the psychology lab, real-life examples of conspiracy theory beliefs driving aggression and physical violence have been reported with alarming regularity. As I mentioned in the previous chapter, in the aftermath of the 2012 Sandy Hook Elementary School shooting that left 20 schoolchildren and 6 staff persons dead, denialists harassed and threatened grieving parents based on conspiracy theories claiming that the shooting never happened, telling them they were "crisis actors" and that their children never existed. In 2016, the Pizzagate conspiracy theory inspired Edgar Maddison Welch to set off on his one-man raid on Comet Ping Pong. In 2020, during the first year of the COVID-19 pandemic, a train engineer intentionally derailed a locomotive to call attention to the conspiracy theory that the Navy hospital ship *USNS Mercy* anchored in the Port of Los Angeles to provide support for the pandemic was there for some other sinister purpose. In Ottawa, a Canadian Armed Forces reservist crashed his pick-up truck loaded with firearms and ammunition through the gate of Rideau Hall to apprehend Prime Minister Justin Trudeau, driven in part by the conspiracy theory that Bill Gates and other powerful global elites had orchestrated COVID-19 during a pandemic preparedness exercise called Event 201. In the United Kingdom, conspiracy theories about 5G networks causing COVID-19 were responsible for arsonists setting some 77 cellphone towers on fire along with several assaults on cellphone carrier employees. Then, in 2021, the conspiracy theory claiming that the US Presidential election had been stolen motivated an angry mob of thousands—some of them equipped with zip-ties with apparent plans to take political leaders hostage—to lay siege to the US Capitol in an act of insurrection that caused $30 million in damages and left countless injured and several dead. Finally, QAnon conspiracy theories were implicated in several headline-grabbing cases of spousal murder and infanticide from 2020 to 2022. While it could be rightly argued that these select few examples of conspiracy theory–related violence represent relatively rare occurrences, there has been cause enough for concern that the FBI declared "conspiracy theory–driven extremists" a growing threat in 2019, with an internal memo stating that "these conspiracy theories very likely will emerge, spread, and evolve in the modern information marketplace, occasionally driving both groups and individual extremists to carry out criminal or violent acts."[27]

Meanwhile, as these incidents and more have unfolded over the past several years, conspiracy theories have been widely covered and promoted by mainstream news outlets so that, regardless of how deeply they're actually believed—but keeping in mind the illusory truth effect that I discussed in the previous chapter—they've breached mainstream awareness and have become ingrained in public consciousness along the way. Political leaders have also publicly endorsed and promoted conspiracy theories to achieve various ends, including successful bids for election, to deflect blame, or to incite political revolt. If there isn't enough evidence to claim that we're living in a Golden Age of Conspiracy Theories based on their prevalence, perhaps there is enough to say that we're living in a Dark Age of Conspiracy Theories based on their consequentiality and potential for harm. With the existential threats of COVID-19, political violence, and climate change hanging over our heads and the fate of the world more than ever before in our lifetimes hinging on our ability to curb conspiracy theory belief, that's not too much of a stretch.

The Psychology of Conspiracy Theory Belief

Why is it that some—and so many—people are drawn to conspiracy theories? One reason is that conspiracy theories provide accounts of events that offer much more compelling drama—just like much of the misinformation that I discussed in Chapters 4 and 5—than the mundane truth. Indeed, Hollywood has been capitalizing on conspiratorial cloak-and-dagger narratives as a recipe for entertainment success for years. The 1962 film *The Manchurian Candidate*, starring Frank Sinatra and Angela Lansbury, was a hit in its day and was turned into a successful reboot with Denzel Washington and Meryl Streep 40 years later. Oliver Stone's *JFK* starring Kevin Costner and Richard Donner's *Conspiracy Theory* with Mel Gibson and Julia Roberts were both 1990s blockbusters, earning hundreds of millions of dollars in the box office to date. And the space alien conspiracy theory TV series *The X-Files* ran for more than a decade, spawning two spinoff feature films and catapulting its actors David Duchovny and Gillian Anderson into superstardom.

The appeal of conspiracy theories is also rooted in the fact that, unlike delusions and misinformation that are by definition false, they're at least plausible accounts with some possibility that they might turn out to be true and cease to be theories. As conspiracy theory believers like to remind us, real-life conspiracies have unquestionably occurred in not-too-distant history, such as the CIA's "mind control" project MK-Ultra or the FBI's COINTELPRO surveillance program, both of which ran from the 1950s into the 1970s. In defending conspiracy theory beliefs, it has therefore been argued that conspiracy theories represent vital democratic mechanisms to push back against unbridled power and abuses of authority. Some even claim that the term "conspiracy theory" was invented by the CIA to delegitimize counternarratives to the Warren Report's conclusions about the assassination of JFK, although, ironically, that itself is an unsubstantiated conspiracy theory.[28]

Over the past couple decades or so, psychology researchers have been examining the popularity of conspiracy theories from the consumer end, trying to determine what might be different about those who are drawn to and believe them. In the process, they've come up with a veritable laundry list of psychological quirks and traits that are associated with either "conspiracy belief" or "conspiracist ideation" evidenced by endorsement of specific conspiracy theories, or more generalized "conspiratorial thinking" or "conspiracy mentality" evidenced by agreement with statements like "many very important things happen in the world, which the public is never informed about"; "even though we live in a democracy, a few people will always run things anyway"; and "much of our lives are being controlled by plots hatched in secret places." To catalog the psychological features of conspiracy theory believers, University of Kent professor of social psychology Karen Douglas and other leading conspiracy theory researchers around the world have grouped them into three main categories of "needs" or "motives" that they call "epistemic," "existential," and "social."[29]

Epistemic motives relate to our need for causal explanations of events, whereas existential needs refer to our need to feel safe. Together, specific epistemic and existential needs for certainty, closure, and control—what I refer to as the "three C's"—tend to be particularly heightened during times of social unrest and societal crises, when explanations for events that are either unsatisfying or lacking leave us feeling threatened.[29] Conspiracy theory belief has also been associated with a "teleologic bias" and "hypersensitive agency detection," meaning a preference and tendency to attribute events to a higher purpose or "ultimate cause."[30] Such psychological needs and cognitive biases help to account for the appeal of conspiracy theories related to the sudden and unexpected deaths of beloved political leaders like JFK or international celebrities like Princess Diana—narratives involving random events such as a lone gunman or an unfortunate car crash fail to alleviate the feelings of horror and helplessness we're left with in the face of tragedy. The same could be said for 9/11, mass shootings, or COVID-19: being told that horrible, life-threatening events simply happen due to chance with no preventable cause is terrifying. Psychologists argue that conspiracy theories might provide us with an alternative account of traumatic events with clearer causes that, in theory, we could do something about. As we might expect, however, explanations involving secret plots and the nefarious intentions of powerful shadow figures controlling the world don't actually quell anxiety or make anyone feel any safer.[31]

Another psychological need associated with conspiracy theory belief is "need for uniqueness."[32] This social motive suggests that conspiracy theories are appealing to some believers because they make them feel as if they've gained privileged access to the "real truth" while the rest of us "sheeple" mill about wearing blinders. As I discussed in Chapters 4 and 5, social media influencers can also make a fortune promoting misinformation and conspiracy theories while attaining celebrity status in the process. At a societal level, social psychologists have found that conspiracy theories can help to preserve a sense of "collective narcissism" by blaming "outgroups" like immigrants and foreign countries for societal crises.[33] Such scapegoating helps to account for the popularity of the COVID-19 "lab leak" conspiracy theory claiming that SARS-CoV-2 was a bioweapon deliberately manufactured by China. Promoted by President Trump and other American political leaders early on during the pandemic, the belief was endorsed by 23–31% of Americans in March of 2020.[34] Playing a tit-for-tat blame game, Chinese officials advanced their own counter-claim that the virus was brought to China from the United States during the Military World Games held in Wuhan months before the pandemic.[35] This conspiracy theory was reinforced by Russian state media and even a few social media influencers here in the United States.[36]

Beyond psychological needs, researchers have also found that conspiracy theory belief is associated with many of the same cognitive quirks that I've addressed throughout this book thus far. These include "faith in intuition" over objective evidence and overconfidence when forming beliefs, as I discussed in Chapters 1 and 2; susceptibility to the "conjunction fallacy," an error in the kind of probabilistic reasoning that I covered in Chapter 2 that involves overestimating the likelihood of two events occurring together; and a relative lack of the same kind of analytical thinking that Pennycook and Rand found can shield us from succumbing to fake news that I described in Chapter 4.[27,37] Motivated reasoning also plays a key role in conspiracy theory belief. On the one hand, the idea that conspiracy theories are appealing to those with a greater conspiracy mentality or a preference for conspiratorial narratives is supported by the finding that belief in one conspiracy theory predicts belief in others, even sometimes when they're contradictory.[12] But, on the other hand, political conspiracy theory beliefs don't tend to cross party lines—those drawn to conspiratorial narratives clearly prefer narratives that are aligned with their political orientation and reject those that oppose it.[38]

It remains a matter of seemingly endless debate and finger-pointing as to whether conservatives or liberals might be more susceptible to conspiracy theory belief. I'll revisit the relationship of political alignment to false beliefs in Chapters 8 and 9, but suffice it to say for now that while some researchers have found evidence of greater conspiracy theory belief among conservatives,[39] this is probably an artifact of the specific conspiracy theories that researchers ask about within a given survey. While there's no doubt that conservatives are more likely to endorse certain conspiracy theories like the "birther" claim that President Obama wasn't born in the United States, it's easy to find examples of conspiracy theory belief on both sides of the political fence if we step back to look objectively at the bigger picture. Indeed, in the wake of the 2016 Presidential election, it seemed as if everyone in the United States believed in one of two conspiracy theories, with liberals rallying behind "Russiagate" that claimed that Trump had colluded with or was even an asset of Vladimir Putin's and conservatives complaining that the real conspiracy could be found in the fictive origins of the Steele dossier and the "witch hunt" that was Robert Mueller's special counsel investigation. Democrats claimed that voter fraud (and Russia's meddling) contributed to Hillary Clinton's defeat in 2016; Republicans claimed it resulted in Trump's loss four years later. Before COVID-19, vaccine conspiracy theories were a justifiable stereotype of affluent, liberal parents living in Marin, California, and have been a central component of longtime Democrat-turned-Independent Robert F. Kennedy Jr.'s campaigning, but now are more likely to be endorsed by conservatives. If these anecdotal examples aren't convincing enough, a recent study by University of Louisville political scientist Adam Enders and his colleagues including Uscinski analyzed the results of 40 surveys conducted in the United States and abroad over the past decade and found that while some individual conspiracy theory beliefs were more common within one or another political orientation, overall, conspiracy theory belief was just as common among liberals as conservatives.[40] Not only that, but the *same* conspiracy theories were equally likely to be believed by partisans when they were manipulated to specify that the conspirators were from the opposing party. These findings make clear that we shouldn't dismiss conspiracy theory beliefs as the paranoid rantings of our political opponents. The reality is that we're all vulnerable to conspiracy theory beliefs when they align with our worldviews.

Mistrust and Misinformation Redux

A few years after 9/11, while lecturing about the gray area between delusions and delusion-like beliefs, I assigned a novel task to two of my medical students. I asked one of them to find evidence to defend the "truther" conspiracy theory that 9/11 was an inside job and the other to find evidence to refute it, and I instructed both of them to search for evidence on YouTube. When they reported back, we held something of a mini-debate on the subject. Martin, who was supposed to defend the truther conspiracy theory, presented information that he'd found about the temperature of ignited jet fuel being insufficient to melt the steel columns of the World Trade Center, sequential explosions consistent with controlled demolition, missiles fired from the hijacked airplanes into the towers and the Pentagon, and the unexplained collapse of Building 7 that was suspiciously reported by the BBC while it was still standing in plain sight in the video feed. Holly, who was supposed to refute those claims, presented a synopsis of the National Institute of Standards and Technology's report that explained how burning jet fuel didn't melt the steel columns but did weaken them sufficiently to cause collapse, revealed that videos of missiles being shot from the hijacked airplanes represented doctored footage, clarified that Building 7 was damaged and set ablaze by falling debris with its inevitable collapse reported prematurely in the news, and debunked claims about Jews not being among the victims of the terrorist attack. After they'd presented their respective arguments, both Martin and Holly stated that they felt more or less convinced by the evidence they'd uncovered, and they each stuck to their guns despite hearing what the other had said. At the end of the day, Martin might not have become a real 9/11 truther, but he demonstrated himself to be at least as much a truther as Kyrie Irving is a flat Earther.

Now, you could argue that the conditions of my teaching exercise with Martin and Holly were too contrived to tell us anything about actual conspiracy theory believers in the real world. After all, they were each assigned which "pro" and "con" stance to defend, just like in my old high school debates that I complained about in Chapter 3. But that assignment wasn't so different from the geographical "accidents of birth" that dictate our cultural beliefs and the other novel ways that we find ourselves inside echo chambers that I discussed in that chapter as well. The assignment also wasn't haphazard—I specifically asked Martin to defend truther claims and Holly to refute them because I had a kind of gut instinct about each of them. Martin seemed more comfortable playing the role of a contrarian or rebel, whereas Holly seemed more "by the book." Maybe I was picking up on Martin's need for uniqueness or, like Kyrie Irving's willingness to "go against the grain," his belief in what some conspiracy theory researchers have referred to as the "malevolence and deceptiveness of officialdom."[12] In any case, he seemed less likely to accept official government narratives at face value.

Whatever the differences in their psychological make-up, the real lesson of Martin and Holly is just how similar they were. They were both excellent medical students equipped with college educations from top universities and solid analytic thinking skills. Both considered themselves politically left-leaning. Both demonstrated confirmation bias and motivated reasoning in their search for evidence as well as in their dismissal of counterevidence that they came across on YouTube and heard from each other during our debate. And while Martin may have had a little contrarian in him, he hardly fit the "conspiracy theorist" stereotype of a "tin foil hat-wearing paranoid crank." At that point in time, he wasn't particularly feeling a need for certainty, control, or closure about 9/11 any more than a flat Earther needs the "three C's" to cope with the roundness of our planet.

These individual realities reveal the limitations of the more generalized psychological research findings about conspiracy theory believers. The psychological needs and cognitive quirks that have been found to be associated with conspiracy theory belief aren't present in all conspiracy theory believers and may be more associated with certain conspiracy theory beliefs than others. And when psychologists tell us that conspiracy theory believers tend to have certain psychological traits, they're talking about quantitative—not qualitative—differences. In other words, we all have the same psychological needs and cognitive quirks that conspiracy theory believers have. Almost all of us have needs for certainty and control, just as most of us like to feel unique or special at times. And most of us have less than perfect analytical thinking skills. So, when studies find that such needs and quirks are associated with conspiracy theory believers, it means that those needs are merely heightened compared to those without conspiracist ideation, often amounting to a relatively small difference.

While those small differences might very well impact who believes in a conspiracy theory or not, too often the psychological research on conspiracy theories is misinterpreted to mean that conspiracy theory believers are afflicted by some all-or-none deficit, like being ignorant, paranoid, gullible, liberal, conservative, or some combination thereof. This deficit claim is frequently used either pejoratively or to suggest that conspiracy theory believers are mentally ill or suffering from some kind of "mass delusion" or "mass psychosis." But, as I hopefully made clear in Chapter 1, delusion-like beliefs like conspiracy theories that might sound outlandish on the surface but are ultimately learned and shared beliefs about the world without a self-referential component shouldn't be conflated with delusions.[41] And while conspiracy theories can certainly be associated with a gamut of negative emotions like anxiety, distress, dysphoria, and anger so that spending time immersed in them often isn't particularly mentally healthy, that's not the same as saying that conspiracy theory believers are mentally ill. Instead, keeping in mind that a substantial majority of respondents to international surveys believes in at least one conspiracy theory,[11] the take-home message of psychological research to date should be that all of us share a universal vulnerability to conspiracy theory belief.

If we are to really appreciate why conspiracy theory beliefs are so pervasive, we therefore need a more normalizing and humanizing framework beyond just thinking about psychological needs and cognitive quirks. The one that I find most helpful is the same framework that I outlined in the previous chapter. That is, belief in conspiracy theories can best be understood as stemming from mistrust that leaves us vulnerable to both misinformation and disinformation.[5] This perspective removes conspiracy theory belief from the narrow and "othering" vacuum of only existing "within the head" and situates it squarely within our social interactions with other people and the information that's out there in the rest of the world. From that vantage point, we can better understand how we've come to be living in a Dark Age of Conspiracy Theories.

Trust No One

Throughout this book so far, I've been making the case that, for good or bad, trust in the testimony of others is a powerful determinant of what we believe. By the same token, mistrust is just as important in determining who and what we choose *not* to believe. The definition of conspiracy theories that I offered earlier highlights that they're part negation of authoritative accounts and part affirmation of conspiratorial counter-narratives. What lies at the root of the negation is what I referred to as "epistemic mistrust" in Chapter 5—that is, a mistrust of authoritative accounts, officialdom, accepted explanations, and conventional wisdom.

There are a myriad of variations and causes of epistemic mistrust, but, for the sake of convenience, they can be grouped into three broad categories. The first lies on a continuum of unwarranted paranoia that can, at the extreme end, be symptomatic of mental illness. For example, in Chapter 3, I wrote about how some individuals with paranoid delusions about being followed, surveilled, harassed, or otherwise persecuted sometimes gravitate to conspiracy theories about more generalized "gang-stalking" for the validation they provide regarding their own subjective experiences. In Chapter 5, I mentioned that "epistemic vigilance" represents a kind of normal skepticism at the healthy end of a paranoid continuum. Somewhere in the middle is a kind of global mistrust that psychologists and psychiatrists refer to as a "paranoid style" that's linked to having a "paranoid personality." People with this personality style often endorse all-or-none cognitive distortions like "everyone lies" and "no one can be trusted" that amount to denialism—that is, a blanket rejection of authoritative sources of information. It should come as no surprise that this kind of subclinical or less-than-delusional worldview has often been found to be associated with conspiracy theory belief since conspiracy theories themselves involve narratives about secret plots and shadowy cabals up to no good. This association is probably tautological, meaning that the reason that conspiracy theory beliefs and paranoia are correlated is that they're just facets of the same construct of mistrust.

The second category of mistrust that can underlie conspiracy theory belief is less a feature of individual psychopathology or personality style and more a characteristic of social groups and their interactions. Conspiracy theories held by one social group about another social group—what psychologists and political scientists refer to as "intergroup conspiracy theories"—often amount to strategies of avoiding responsibility for societal crises and deflecting blame onto "enemies from within," marginalized outgroups, or foreign adversaries, often in line with long-standing political conflicts and racist or xenophobic attitudes. As I suggested earlier, when a group's collective narcissism and social integrity are perceived to be threatened in times of societal upheaval, needs for certainty, control, and closure are heightened, and fears can be assuaged by identifying other social groups as the source of that threat.[42]

The use of intergroup conspiracy theories for scapegoating in this way often occurs without the conscious awareness of believers despite being employed as deliberate propaganda by political leaders. As I noted previously, "lab leak" conspiracy theories about COVID-19 being deliberately manufactured as a bioweapon, just like labeling it the "Chinese flu," served the purpose of shifting national responsibility for the mismanagement of the pandemic onto foreign adversaries. Ever since Russia invaded the Ukraine in 2022, Putin has justified the war based on the conspiratorial claim that the country is rife with Nazis backed by the West who are intent on genocide through biological warfare. In much the same way, conspiracy theories have been used to malign Jews for millennia, with modern-day versions implicating the Rothschilds, George Soros, and even "Jewish space lasers" as part of a New World Order perpetuating the same tropes that were laid out in the antisemitic *Protocols of the Elders of Zion* a century before, as well as "blood libel" conspiracy theories dating back to the third century. The relationship between mistrust and intergroup conspiracy theories is reciprocal, meaning that mistrust of outgroups can lead to conspiracy theory beliefs that in turn fuel more mistrust, resulting in greater prejudice toward and discrimination against immigrants and ethnic minority groups.[43]

Finally, epistemic mistrust can sometimes be completely rational and justified due to real-life trust violations. Here in the United States, conspiracy theory beliefs about AIDS—such as that the HIV virus was created and deliberately spread among marginalized communities by the CIA—are overrepresented among African Americans.[44] The underlying cultural mistrust that has provided fertile soil for such beliefs has been attributed to a long history of trust violations perpetrated by the American medical establishment against Black people. Such violations include the 1932–1979 Tuskegee Syphilis Study that withheld proper medical care for Black Americans in the name of science, but also a much longer list of systemic ethical breaches dating back to colonial times that have been detailed in Harriet Washington's 2007 book *American Apartheid*.[45] Keeping this example in mind, it should come as no surprise that conspiracy theory belief tends to be more frequent in countries where there's objective evidence of corruption at the societal level.[46]

Four years after the start of the pandemic, earned mistrust has contributed to the fact that conspiracy theories about COVID-19 being a hoax and vaccines being a Big Pharma scam continue to flourish. During the early months of COVID-19, genuine uncertainty about the novel viral outbreak was coupled with misstatements and possibly even deliberate misdirection by the World Health Organization, the Centers for Disease Control and Prevention (CDC), and the US Surgeon General who told us not to wear masks, perhaps so that they would remain available to frontline healthcare workers. When the CDC changed its stance a few months later, the damage to its credibility was already done in the eyes of many whose mistrust led them straight into the waiting arms of conspiracy theory counter-narratives. Other early misstatements about the SARS-CoV-2 virus not being spread by airborne transmission that were later walked back didn't help. Neither did National Institute of Allergy and Infectious Disease Director Anthony Fauci's admission at the end of 2020 that he "moved the goalposts" on the percentage of Americans needed to attain herd immunity or the CDC's suppression of an internal document that was later leaked to reveal that "breakthrough" infections and transmission of the SARS-CoV-2 delta variant were indeed occurring among the vaccinated.[47]

For consumers desperate for reliable information during a crisis, the trustworthiness of institutions of epistemic authority must come before trust. Lack of transparency and orchestrated misdirection, even if well-intentioned for some greater good, is a surefire way to erode trust and open the door to belief in conspiracy theories and other types of misinformation. Experts and institutions of epistemic authority need to do a better job of owning the earned mistrust that lies at the heart of conspiracy theory beliefs today.

***

The connection between conspiracy theory belief and mistrust of officialdom and authoritative accounts of information has been supported across a number of studies, but, just as epistemic mistrust can have many causes, it can also have many different targets so that exactly who one mistrusts can predict what kind of specific conspiracy theory one might find appealing.[5] As we might expect then, belief in political conspiracy theories that's associated with mistrust in government and political leaders occurs within the boundaries of partisan-motivated reasoning. Although overall conspiracy theory belief is just as common on both sides of the political fence, as discussed previously, there is evidence that conspiracy theories are more likely to be embraced by "losers" who have become dispossessed of political power.[13] Other research has found evidence of a "president in power effect" whereby individuals have less trust in government and more belief in political conspiracy theories when the sitting president is a member of one's opposing political party.[48] The largest survey of conspiracy theory belief performed to date across 26 countries likewise confirmed that conspiracy mentality tends to be stronger among voters whose political parties aren't in power, but also among the extremes—that is, the far-right and far-left—of political orientation.[49]

Another more enduring finding than any left–right political asymmetry has been an association between conspiracy theory belief and populism. In recent years and across several studies and polls, populist attitudes and beliefs that transcend left–right politics have been associated with greater belief in contemporary conspiracy theories related to COVID-19, vaccines, and climate change.[50] As with the likely tautological link with paranoia, the connection between conspiracy theory belief and populism probably reflects that they're different manifestations of the same attitude. After all, populism is a political movement defined by the opposition of "the people" against "elites" in power, typically framed as a battle between good and evil—the same Manichean dynamic portrayed within conspiracy theory narratives. Uscinski and Enders take this connection one step further, linking conspiracist ideation, populism, and a preference for Manichean narratives all under a broader heading of what they call "anti-establishment sentiment."[51]

Historically, populist movements and anti-establishment attitudes have emerged within both liberal and conservative political parties. Likewise, liberals and conservatives are equally vulnerable to motivated reasoning about science, with mistrust of scientists depending on what exactly they're saying.[52] If populist attitudes promoting mistrust in scientific expertise have become more tightly entwined with conservative and right-wing politics in recent decades, it's likely because scientific consensus around certain issues has created "inconvenient truths"—as former US Vice President Al Gore put it back in 2006—that run counter to conservative worldviews, concerns, and political agendas.[14,53] This, more than any asymmetric political vulnerability to conspiracy theory belief, helps to explain why it's primarily conservatives that are disputing consequential issues of scientific authority such as climate change, COVID-19, and the efficacy and safety of vaccines in the political arena today.

Taking a step back from politics for now—I'll return to the topics of America's political divide in Chapter 8 and the political asymmetry of belief in climate change in Chapter 9—to summarize the bigger picture, it's safe to say that conspiracy belief is rooted in epistemic mistrust, at both the level of the individual and across society. Mistrust is the constant that runs between Kyrie Irving, my student Martin, and all those who refute voices of authority with conspiratorial counter-narratives. It's therefore no coincidence that the Dark Age of Conspiracy Theories, in which conspiracy theory beliefs have come to bear on public policymaking, coincides with a modern era of pervasive and declining mistrust in institutions of epistemic authority. A 2018 YouGov poll found that 52–85% of Americans and Europeans believed in at least one conspiracy theory and that 51–84% mistrusted journalists, 66–92% mistrusted senior officials in the United States, 52–83% mistrusted senior officials of the European Union, 42–83% mistrusted people who run large companies, and 27–66% mistrusted "people you meet in general."[20] The erosion of trust, reflecting a growing allegiance to the populist revolt against the wisdom of elites, political leaders, and scientific expertise has taken us down a path that converges on Uscinski and Enders's "anti-establishment sentiment," Isaac Asimov's "anti-intellectualism," RAND researchers Jennifer Kavanaugh and Michael Rich's "truth decay," Tom Nichols's "death of expertise," and Stephan Lewandowsky's "post-truth world." To that list of societal ills attributable to epistemic mistrust, we can now add the Dark Age of Conspiracy Theories.

Do Your Own Research

In the *X-Files* TV series, the protagonists Fox Mulder and Dana Scully were FBI agents who found themselves falling down the rabbit hole of conspiracy theories and seemingly paranormal phenomena during each episode. Like my students Martin and Holly, Mulder played the part of the conspiracy theory believer while Scully played the part of his scientifically skeptical foil. In Mulder's basement office, where he'd been banished for his unconventional work, he proudly displayed a poster with the grainy photo of a flying saucer and the bold print words, "I want to believe."

I mentioned earlier that mistrust and conspiracy theory belief are reciprocal, with one fueling the other. As I suggested in the previous chapter, the idea that conspiracy theory believers like Fox Mulder want and need to believe something is a key point in this dynamic that explains why people who mistrust authoritative sources of information don't just end up epistemic nihilists who don't believe anything. For them, conspiracy theories offer particularly appealing counter-narratives that validate their worldview of elites and voices of authority as bad actors and liars. Research on the spread of conspiracy theories within online discussion forums tells us that it's this drive to believe that leads people down the rabbit hole in a search of such counter-narratives.[54] As with the search for information more generally, "just asking questions" and "doing one's own research" amounts to an active process guided by both confirmation bias on steroids as well as motivated reasoning, as I discussed in Chapters 3 and 4. Because this active search for explanations often occurs across a modern media landscape where misinformation and disinformation are ubiquitous, the familiar term "conspiracy theorist" often isn't accurate. Those who believe in conspiracy theories aren't usually theorizing so much as they're assimilating conspiracy theory narratives and synthesizing them into the "evidence" that's used to justify their mistrustful worldview.[5] This characterization highlights the crucial difference between scientific research that actually tests hypotheses in a controlled fashion and the process of "doing your own research" that amounts to reinforcing preexisting suspicions about the untrustworthiness of authoritative sources of information through confirmation bias and motivated reasoning.

Within the flea market of opinion that I described in Chapter 4, online media sources are key players in reinforcing a conspiratorial worldview. When Texas Tech University psychologist Asheley Landrum and her research team attended the 2017 Flat Earth International Conference, attendees told them that they had "only come to believe the Earth was flat after watching videos about it on YouTube."[55] Marc Sargent, a prominent social media influencer who makes a living promoting flat Earth conspiracy theories, has likewise claimed that, without YouTube, the modern flat Earth movement "wouldn't exist."[3] Before being deprioritized by that social media platform, Sargent's YouTube videos, just like Eric Dubay's infamous "200 Proofs Earth Is Not a Spinning Ball," attracted viewers numbering in the millions. Kyrie Irving was presumably one of them.

Beyond these anecdotes, a 2007 study found that belief in 9/11 conspiracy theories was significantly associated with being a consumer of grocery store tabloids and online blogs as opposed to more "legitimate media sources."[56] In the past few years, the evidence that online informational sources and social media contribute to conspiracy theory belief has become even more clear. A 2019 YouGov survey found that those endorsing populist attitudes were more likely to get their news from social media sources like Facebook, Twitter, YouTube, and WhatsApp and were roughly twice as likely to believe in conspiracy theories.[57] Political scientists Dominik Stecula and Mark Pickup likewise reported survey results revealing that the use of YouTube and Facebook for news was associated with a greater likelihood of conspiracy theory belief.[58] In a separate study, they also determined that the association between populist attitudes and conspiracy theory belief was stronger among those who consumed conservative media sources for news.[59] During the pandemic, results from an international survey also indicated that belief in conspiracy theories and misinformation about COVID-19 was related to both exposure to and trust in digital media sources over traditional media sources like newspapers, radio, and television.[60] Even Uscinski and Enders, who emphasize that social media doesn't deserve all of the blame for conspiracy theory belief, found that in their own 2020 survey those who got their news from social media were more likely to endorse belief in conspiracy theories.[61]

While people believed in conspiracy theories long before social media existed, the term "do your own research" as it's used today largely refers to searching for information online. But Uscinski and Enders are right that it's not as if most people are randomly and unwittingly falling down conspiracy theory rabbit holes on the internet. As I explained earlier, coming to embrace conspiracy theory beliefs typically involves an active search by those with epistemic mistrust and a conspiracy mentality. For those mistrustful of authoritative sources of information, the misinformation that's out there in the world justifies, reinforces, and caters to that attitude. If we are living in a Dark Age of Conspiracy Theories, it's not only because mistrust is so common or because misinformation is so plentiful—it's also that both mistrust and misinformation are ubiquitous, with one feeding into the other.

Follow the Money

Back in the early 2000s, I treated a young man admitted to the hospital in a frazzled and agitated state who was coming down off a methamphetamine high. He paced back and forth down the hallways, demanding release from his involuntary psychiatry hold and complaining that he felt unsafe on the unit because he was being targeted by the government because of "the things I know." When I inquired just what it was exactly that he knew, he offered a scattered account of being surveilled due to "psy-ops" that were part of a "government cover-up." After the self-referential part of his acute paranoia subsided with a few days of sobriety, he gave up the claim that he was in any personal danger but continued to go on about how all Americans were imperiled by the "New World Order." When he could see that I was skeptical, he told me, "check out InfoWars and you'll see the truth" while making the "sign of the horns" with both hands. While many of the hospital staff assumed he was still delusional, I took his advice and checked out InfoWars—probably the first time I'd ever done so—and promptly realized that he was just parroting some guy named Alex Jones before anyone knew who that was. I discharged him from the hospital.

Uscinski and Enders like to remind us that, beyond the role of social media, "public opinion formation is substantially a top-down affair."[62] In other words, as I argued in Chapter 5, it can't be overemphasized just how influential the apex predators of the disinformation food chain are when they disseminate and encourage belief in conspiracy theories based on their own political agendas and vested interests. And yet, the irony of conspiracy theories is that, all too often, believers are so blinded by epistemic mistrust that they fall victim to disinformation without seeing the real conspiracy hiding right under their noses. In defending conspiracy theory beliefs, it's often advised that one "follow the money," without considering who might be funding and profiting—either financially or politically—from disseminating conspiracy theory counternarratives.

In the political arena, conspiracy theories are often employed as a tried and true tactic of casting aspersions on adversaries, as Trump demonstrated in 2014 by promoting the Obama "birther" conspiracy and in 2016 by claiming that Texas Senator Ted Cruz's father was an associate of JFK's assassin Lee Harvey Oswald. Conspiracy theories can also serve as a convenient strategy for political leaders to divert public attention, as illustrated by Trump's promotion of the "lab leak" conspiracy theory about COVID-19. In 2020, when Trump used the word "hoax" to describe Democrats' criticism of his handling of the pandemic, many liberals took it to mean that he was claiming that COVID-19 itself was a hoax. While that interpretation probably didn't reflect the true intent of his words,[63] the fact that, over a year later there was a 70–16% split over the conspiracy theory that the threat of COVID-19 was exaggerated for political reasons, with conservatives four times as likely to believe it than liberals, suggests that even an ambiguous statement by Trump was enough to be taken to heart by his supporters.[18] As political scientists Nancy Rosenblum and Russell Muirhead argue in their 2019 book *At Lot of People Are Saying: The New Conspiracism and the Assault on Democracy*,[64] the use of innuendo (remember the "innuendo effect" that I mentioned in the previous chapter) in the post-truth era and the common conspiracy theory believer's refrain of "just asking questions" amounts to a kind of "conspiracy without the theory" that's often more than adequate to encourage conspiracy theory belief when it's handed down from the top of the disinformation food chain.

Returning to the kind of examples of apex predators profiting financially from disinformation that I outlined in the previous chapter, modern conspiracy theories about vaccines causing autism were born out of fraudulent research claims made by Andrew Wakefield, a one-time gastroenterologist who lost his medical license as a result. Investigative journalism later revealed that, in addition to research fraud, Wakefield had substantial conflicts of interest including being paid handsomely for funneling patients to a law firm handling a class action lawsuit against a vaccine manufacturer after he'd applied for a patent to develop his own competing vaccine.[65] Despite this scandal being labeled "the most damaging medical hoax in 100 years,"[66] Wakefield went on to achieve Messiah status in the anti-vaccine movement and now makes a living as one of its top spokespersons, peddling anti-vaccine conspiracy theories and telling eager audiences not to trust doctors and pharmaceutical companies.

Similar investigative journalism exposed another timely object lesson that unfolded during the pandemic. In July 2020, Trump retweeted an online video featuring a group of physicians dressed in white coats calling themselves America's Frontline Doctors (AFD) who stood behind a podium on the steps of the US Supreme Court to "tell the truth" about COVID-19. In addition to making a variety of other false claims about COVID-19 being less deadly than the flu and masks not being useful to prevent its spread, they touted the effectiveness of hydroxychloroquine (a medication typically used to treat malaria and lupus) as a "cure" and called studies that found it to be ineffective "fake science" sponsored by "fake pharma companies."[67] The video went viral on social media and sparked a deluge of false claims about hydroxychloroquine's effectiveness for treating COVID-19 on Twitter, with 84% of the 2.7 million tweets about the drug found to contain misinformation in just nine days following Trump's retweet.[68]

Reporting by *The Washington Spectator* later revealed that AFD was part of a "highly orchestrated effort" by the Council for National Policy, a "coalition that coordinates initiatives among conservative megadonors, political operatives, and media owners" that activated a group of "extremely pro-Trump" doctors who could help exploit Americans' trust of physicians in order to downplay concerns about COVID-19 and advocate for reopening the country and ending the lockdown.[69] Throughout 2021, AFD continued to spread anti-vaccine disinformation and conspiracy theories while adding an endorsement for ivermectin, an anti-parasitic medication used more often in veterinary medicine to treat livestock than in human medicine, as another purportedly "safe and effective" medication to treat COVID-19. By the end of August 2021, *Time* magazine reported that AFD's founder, Simone Gold, an emergency room physician who was charged and convicted for her participation in the US Capitol insurrection, had launched a nationwide RV tour earlier in the year offering VIP "meet and greet" tickets at $1,000 a pop but was canceling many of her scheduled appearances without advance notice.[70] Meanwhile, AFD transformed itself into a nationwide telemedicine practice offering hydroxychloroquine and ivermectin prescriptions for a $90 consultation fee while leaving hundreds of customers in a lurch after failing to deliver the medications. In September 2021, *The Intercept* revealed leaked data demonstrating that AFD collected nearly $7 million in consultation fees over just two months and that, in partnership with Ravkoo, an online pharmaceutical service charging inflated prices for generic medications including not only ivermectin and hydroxychloroquine but also zinc and vitamin C, AFD profited from an additional $8.5 million in drug sales between November 2020 and September 2021.[71] By the end of 2022, after serving out most of her 60-day prison sentence, Gold found herself embroiled in a bitter lawsuit by AFD's board of directors alleging that she'd used company funds on a $100,000 private jet trip, a $3.6 million Florida mansion, and personal expenses totaling $50,000 a month.[72]

Finally, on a much larger scale, the claim that climate change is a hoax perpetrated as a part of some globalist conspiracy has been promoted for years by various actors with close financial ties to the fossil fuel industry. The Koch brothers, whose multibillion dollar company Koch Industries is involved in oil refinement and distribution, among other ventures, reportedly spent $127 million over two decades to fund climate change denialism groups.[73] The Exxon Corporation has likewise invested millions of dollars in disinformation campaigns promoting climate change conspiracy theories while suppressing its own internal research from as far back as the 1970s demonstrating that anthropogenic climate change is a reality.[74] US Senator James Inhofe, who wrote a 2012 book about climate change entitled *The Greatest Hoax: How the Global Warming Conspiracy Threatens Your Future* and infamously brought a snowball to the Senate floor during the winter of 2015 to demonstrate that global warming was a myth, has received millions of dollars in campaign donations from the fossil fuel industry throughout his career (I'll return to the topic of climate change denial in Chapter 9).[75]

These few accounts of the real conspiracies behind conspiracy theories—revealed by investigative journalism, not "conspiracy theorists"—echo the examples from Chapter 5 in which apex predators sitting atop the disinformation food chain both exploit and cultivate mistrust by attacking inconvenient truths and replacing them with falsehoods. When conspiracy theory believers tell us to "do your own research," all too often what they're really advising is to do as they did—jump down the rabbit hole to search for and locate the disinformation that's been deliberately put out there in the world by people who profit from it. When we're mistrustful of officialdom and looking for other answers, that kind of disinformation has a way of falling into our laps.

Where We Go One, We Go All

One of the most iconic scenes captured on video from the US Capitol on January 6, 2021, involved a group of rioters who'd entered the building advancing up the steps while a lone Capitol Police officer, Eugene Goodman, backpedaled in retreat. Leading the throng was a man sporting a beard, a beanie, and a t-shirt with a large "Q" and the words "Trust The Plan" on its front who was later identified as one Douglas Jensen. At some point during the day, Jensen posted a selfie video to social media in which he bragged, "this is me, touching the fucking White House ... this is why we're here." His confusion about where he was notwithstanding, he was arrested two days later and, by the following Monday, was indicted on seven federal charges including "violent entry and disorderly conduct in a Capitol building" and "obstructing a law enforcement officer during a civil disorder."

According to an FBI report, Jensen originally "stated that he intentionally positioned himself to be among the first people inside the United States Capitol because he was wearing his 'Q' t-shirt and he wanted to have his t-shirt seen on video so that

'Q' could 'get the credit.' "[76] He went on to admit that he'd "followed QAnon for four years, eventually spending most of his waking, nonworking hours pursuing it and becoming a 'digital soldier' and 'religious adherent.' "[77] But, six months later, his defense attorney claimed that Jensen felt "deceived, recognizing that he bought into a pack of lies" and "became a victim of numerous conspiracy theories that were being fed to him over the internet by a number of very clever people," adding that Jensen became a "true believer" in the QAnon conspiracy theory "for reasons he does not even understand today."[78] Soon thereafter, Jensen was granted a pretrial release from jail with home confinement on the condition that he would stay out of DC, avoid contact with anyone else involved with the January 6 insurrection, not possess a firearm, refrain from alcohol use, and not access the internet or use any internet-capable devices. However, just two weeks after being released, a court supervision officer caught Jensen in his garage watching a "cyber symposium" by My Pillow CEO Mike Lindell that was perpetuating the conspiracy theory that the 2020 election had been stolen by voter fraud. Jensen was taken back to jail, with his attorneys likening his inability to comply with the conditions of his release to "an addiction."[77]

QAnon can be briefly summarized as a political movement based around a core conspiracy claim that Democratic leaders are part of the so-called Deep State, a worldwide cabal of Satan-worshipping pedophiles operating an international child sex-trafficking ring and aspiring to world domination, with President Trump ordained as the savior chosen to deliver us from this evil. This belief system was born out of the Pizzagate conspiracy theory while also resurrecting more long-standing conspiracy theory tropes about a globalist New World Order intertwined with strong undercurrents of antisemitism and Christian evangelicalism. In the fall of 2017, QAnon crystallized into a movement with a new name when an anonymous individual calling himself "Q" and claiming to be a "high level government official" with access to top-secret information began posting on the website 4chan. Q's modus operandi involved cryptic messages that came to be known as "Q-drops" or "breadcrumbs" left for followers calling themselves "bakers," "Anons," and "Q-patriots" to decipher that provided the clues, like pieces of some elaborate puzzle, outlining the path to Trump's victory over the Deep State as part of a coming "Great Awakening" and "Storm." "The Storm Is Coming" and "Where We Go One, We Go All" became popular slogans for adherents, who began to show up at Trump rallies with "Q" signs and t-shirts. This in turn sparked mainstream media coverage so that QAnon became a household word in 2020 and spread internationally during the pandemic, becoming aligned with populist politics and a more general revolt against government lockdowns and mandates for mask wearing and vaccination.

Just how many people have actually come to believe QAnon conspiracy dogma isn't all that clear. In March 2020, polls indicated that only 24% of the US population had heard or read about QAnon, but that percentage had grown to nearly 50% by September.[79] Following Trump's defeat at the end of the year and leading up to the US Capitol insurrection, 39% of Americans endorsed the belief that there was a "deep state" working to undermine Trump, and 17% believed the core QAnon tenet that "a group of Satan-worshipping elites who run a child sex ring are trying to control our politics and media."[80] But a closer look at those poll numbers reveals that such responses reflected a spectrum of belief conviction, with most expressing—as with Kyrie Irving's endorsement of flat Earth conspiracy theories and my student Martin's allegiance to 9/11 truther claims—a more generalized sympathy with QAnon sentiments rather actual die-hard belief. True believers like Jensen, who when questioned by FBI agents asked if the mass arrests of corrupt politicians that Q foretold had taken place yet, are a much rarer bird. In this sense, we might compare QAnon belief to belief in the Bible, with both literalists and figurativists standing under the umbrella of adherents.

Indeed, QAnon has been likened to both a new religious movement and a cult and it's easy to see how its Manichean narrative of an apocalyptic battle between Trump on one side as a Messianic savior and the Satanic Deep State representing evil on the other might appeal to the likes of Trump supporters and Christian evangelicals.[81] But as with conspiracy theories more generally, it would be a mistake to only think of QAnon as the movement of a right-wing fringe. In fact, QAnon has managed to resonate across a variety of ideological groups, including not only Evangelicals and Trump supporters, but populists, libertarians, and sovereign citizens like Robert Drummond that I described in Chapter 5; "anti-vaxxers" and "anti-maskers" incited by the pandemic; those concerned about child trafficking that found QAnon through the SaveTheChildren hashtag on social media; those interested in "wellness," as I'll discuss in the next chapter; and those in other countries who aren't all that concerned with American politics. What's more, QAnon isn't only a political ideology or a conspiracy theory or a religion; with its "breadcrumbs" and "bakers" interacting in online virtual space, it's both a potentially addicting alternate-reality, multiplayer, role-playing game as well as an online social community of like-minded people with anti-establishment sentiments. With all these "hooks" reaching out in different directions, QAnon offers a little bit of bait to almost everyone who's hungry.[82]

We should keep in mind that people sometimes express allegiance with a conspiracy theory belief, especially when responding to polls, without being able to explain the evidence for or against that belief. This was demonstrated clearly in a 2020 survey where some respondents endorsed support for QAnon, but were often unaware of its more outrageous claims and rated specific QAnon-related claims as "true" despite encountering them for the first time.[83] For example, 22% of respondents believed that "a global network tortures and sexually abuses children in Satanic rituals" and 18% believed that "Trump [was] secretly preparing a mass arrest of government officials and celebrities," but those percentages dropped to only 12% and 6%, respectively, after subtracting out those who only heard about those claims for the first time when asked about them during the survey. This makes clear that when people are queried about conspiracy theory beliefs in polls, the details often aren't known or don't really matter. When Trump himself was asked about QAnon ahead of the 2020 election, he replied, "they like me very much. . . . I heard that these are people who love our country"[84] and later added, "I know nothing about QAnon . . . what I do hear about it is they are very strongly against pedophilia and I agree with that ... I do agree with that."[85] If all people know about QAnon is that it's about fighting pedophilia, who going to denounce it in a poll?

* * *

Some version of the quotation "insanity is a sane response to an insane society" has been variably attributed to the sociologist Emile Durkheim or to the psychiatrists R. D. Laing and Thomas Szasz. While this popular aphorism falls short of accounting for the biological underpinnings of pathological delusions that are part of mental illnesses like schizophrenia, it's a fair take on conspiracy theory beliefs, which are much less evidence of individual psychopathology than they are a symptom of a sick society plagued by the twin maladies of mistrust and misinformation.[41]

Having written about conspiracy theories and QAnon over the past several years, including the reasons why people might fall down the QAnon rabbit hole and not want to climb back out, I've been contacted by many family members and other loved ones of "Anons" like Jensen. I've read their stories, followed others like them in the news, and have even provided legal consultation for the defense of one die-hard supporter who was charged with making threats in connection to US Capitol insurrection. If there's a common anecdotal thread running through all the narratives I've heard about QAnon adherents—and especially the literalist true believers who are well-versed in the details of QAnon dogma—it's that they spent increasing time on the internet leading up to the 2020 election, whether laid off from work or stuck at home during the pandemic lockdowns, where they got sucked deeper into the online world of QAnon and farther away from their previous lives and relationships. Jensen's lawyer explained his client's descent in this way:

> Maybe it was mid-life crisis, the pandemic, or perhaps the message just seemed to elevate him from his ordinary life to an exalted status with an honorable goal. In any event, he fell victim to this barrage of internet sourced info and came to the Capitol, at the direction of the President of the United States, to demonstrate that he was a "true patriot."[86]

The Dark Age of Conspiracy Theories shouldn't be written off as the result of some long-standing "paranoid fringe" becoming more unhinged. Each of us shares the potential to believe in conspiracy theories—what has changed in recent years is that all the other social and societal factors that activate our psychological vulnerability to conspiracy theory belief have converged in a kind of perfect storm. A global crisis heightening psychological needs for control, certainty, and uniqueness. Mistrust of authoritative sources of information leaving people vulnerable to widespread misinformation and disinformation. Salvation from forced social isolation among likeminded individuals with anti-establishment worldviews gathered in communal spaces online. And an attraction to an epic battle between good and evil without seeing how that conspiracy theory narrative was exploited by political leaders to incite an angry mob.

Mentally healthy societies don't make pivotal decisions based on false beliefs and conspiracy theories. They don't allow the fabric of democracy to be torn apart or people the world over to die because of conspiracy theory beliefs. And yet here we are.

Where we go one, we go all indeed.

7
Falling for Bullshit

> One man's bullshit is another man's catechism.
>
> —Neil Post

A Case of Bullshit

When I work as a forensic consultant and expert witness for legal cases that involve delusions and delusion-like beliefs, I don't think of my job as helping convict people of crimes or to get them off—that's for the prosecution and defense attorneys to hash out. My role, regardless of which side has hired me, is to determine what people believe, why they believe it, and how that influences their behavior and then explain it all to a jury. And so, when I took the stand to testify about Robert Drummond, the sovereign citizen that I described at the start of Chapter 5, my goal was to characterize the conspiracy theories that he believed and—since he wasn't delusional or otherwise mentally ill—to outline three reasons that he'd come to regard such falsehoods as truth.

The first reason, as I discussed in the past two chapters, was a combination of mistrust and misinformation, which in Mr. Drummond's case meant a resentful distrust of the federal government that had led him to literally buy into a variety of bogus sources through the years that were peddling sovereign citizen dogma. The second reason was that he had more than his share of "conspiracy mentality" such that he was more susceptible to believing conspiracy theories than the average person, as reflected in his agreement with items from a questionnaire called the Generic Conspiracist Belief Scale such as:

- A small, secret group of people is responsible for making all major world decisions, such as going to war.
- The government permits or perpetrates acts of terrorism on its own soil, disguising its involvement.
- Technology with mind-control capacities is used on people without their knowledge.
- Evidence of alien contact is being concealed from the public.[1]

And the third reason, I explained after apologizing to the judge for my language, was that, in addition to conspiracy theories, Mr. Drummond was also more likely than the average person to believe in bullshit.

That's right, I told the jury with a straight face, this was a case of bullshit.

Bullshit, Bullshitters, and Bullshittees

Over the past few decades, bullshit has gone from being a mere curse word and colloquialism to a phenomenon of serious psychological investigation. In this chapter, building upon the previous few that have been making the case for mistrust, misinformation, and motivated reasoning lying at the heart of false belief, I'll explore bullshit as a specific type of rhetoric that's intended to impress us in the service of cultivating unwarranted credibility and trust. As we'll see, although some of us might be more susceptible to it than others, bullshit can often prove quite convincing and therefore effective at steering us away from more reliable sources of information. Consequently, bullshit in its various forms is so pervasive—whether in pseudoscience, postmodern philosophy, or politics—that it's almost impossible not to find ourselves wading knee deep in it from time to time.

The birth of bullshit as a legitimate subject of academic inquiry can be traced back to these words from Princeton philosopher Harry Frankfurt's 1986 essay, simply titled *On Bullshit*:

> One of the most salient features of our culture is that there is so much bullshit. Everyone knows this. Each of us contributes his share. But we tend to take the situation for granted. Most people are rather confident of their ability to recognize bullshit and to avoid being taken in by it. So the phenomenon has not aroused much deliberate concern, or attracted much sustained inquiry. In consequence, we have no clear understanding of what bullshit is, why there is so much of it, or what functions it serves. And we lack a conscientiously developed appreciation of what it means to us. In other words, we have no theory. I propose to begin the development of a theoretical understanding of bullshit, mainly by providing some tentative and exploratory philosophical analysis.[2]

Thirty-five years of academic study later, psychologists have now put together a working conceptual theory of bullshit. To start, we can define bullshit in technical terms as "communications that result from little to no concern for truth, evidence and/or established semantic, logical, systemic, or empirical knowledge."[3] Put more simply, bullshit has been described as "something that implies but does not contain adequate meaning or truth."[4]

Bullshitting isn't just spewing nonsense or gibberish. It's a form of deliberate discourse that's intended to sound meaningful but, on closer examination, isn't actually meaningful. And bullshit isn't the same as lying. A liar knows the truth but fabricates statements to sell people on falsehoods. Bullshitters, by contrast, aren't so much concerned about what's true or not as they're hoping to appear as if they know what they're talking about to "impress rather than to inform."[4] In that sense, bullshitting can be thought of as a verbal demonstration of the Dunning-Kruger effect that I discussed in Chapter 2: when people speak from a position of disproportionate confidence about their knowledge relative to what little they actually know, bullshit is often the result.

In his 2018 study on bullshit, Wake Forest University psychologist John Petrocelli found evidence to support that, beyond the Dunning-Kruger effect, bullshitting tends to happen when there's social pressure to provide an opinion as well as a "social pass" that allows someone to get away with bullshit as a response.[3] In Frankfurt's 1986 essay, as well as his 2005 bestselling book of the same name,[5] he argued that bullshit was so prevalent at the time because the conditions that Petrocelli outlined were key features of an American culture in which people felt entitled if not obligated to offer opinions about everything and where objective reality was often denied in favor of voicing impassioned personal opinions. In today's post-truth world where opinions are routinely conflated with news, facts and expertise have been pronounced dead, and objective evidence is refuted at every turn, as I've described over the past few chapters, it's tempting to claim that the conditions that give rise to bullshit have never been so fertile. Indeed, education wonks Alison MacKenzie and Ibrar Bhatt from Queen's University in Belfast argued just that in 2018.

> [A]long with a pervasive and balkanized social media ecosystem and high internet immersion, public life provides abundant opportunities to bullshit and lie on a scale we could have scarcely credited 30 years ago.[6]

Much like conspiracy theories then, while it's debatable whether bullshit—along with our collective appetite for it within the junk food diet of information that I described in Chapter 4—has really reached epic proportions compared to years past, it would be hard to argue that we don't at least have easier access to a much wider menu of bullshit now than ever before.

***

Every year since 1991, the satirical Ig Nobel Prize has been awarded to honor "achievements that first make people laugh, and then make them think."[7] Past recipients have included actual Nobel Prize winner and physicist Andre Geim for "using magnets to levitate a frog," chemist Mayu Yamamoto for "extracting vanilla flavoring from cow dung," and psychologists David Dunning and Justin Kruger for demonstrating their eponymous effect. But awardees have also run the tongue-in-cheek gamut to include the likes of President of France Jacques Chirac for "commemorating the fiftieth anniversary of Hiroshima with atomic bomb tests in the Pacific" and Vice President Dan Quayle for "demonstrating, better than anyone else, the need for science education." In 2016, the Ig Nobel Peace Prize went to Gordon Pennycook, whose research on fake news I cited in Chapters 4 and 5, along with his co-investigators at the University of

Waterloo and Sheridan College "for their scholarly study called 'On the Reception and Detection of Pseudo-Profound Bullshit.'"

Indeed, while Frankfurt may have given birth to the academic study of bullshit and bullshitting, few have advanced our knowledge of "bullshittees"—those of us who consume and are vulnerable to bullshit—more than Pennycook and his colleagues who developed a questionnaire to quantify susceptibility to a particular kind of bullshit called "pseudo-profound bullshit" that they define as "seemingly impressive assertions that are presented as true and meaningful but are actually vacuous."[4] Pennycook's Bullshit Receptivity Scale (BRS) asks respondents to rate the profoundness of 10 statements constructed from a humorous online site, wisdomofchopra.com, that produces a random but grammatically and syntactically correct mash-up of words taken from actual tweets by Deepak Chopra along with similar "profound-sounding words" from another website called The New Age Bullshit Generator. These include statements like

- Hidden meaning transforms unparalleled abstract beauty.
- Perceptual reality transcends subtle truth.
- Consciousness is the growth of coherence, and of us.
- The future explains irrational facts.
- Today, science tells us that the essence of nature is joy.

Based on tallied responses to the BRS, Pennycook and his colleagues have demonstrated that pseudo-profound bullshit receptivity varies across individuals as a continuous and quantifiable psychological trait. In other words, just like many of the other traits and cognitive biases I've discussed in previous chapters, we're all susceptible to pseudo-profound bullshit in varying degrees. It isn't a matter of "all" or "none"—it's a matter of "how much."

When I told the jury that Mr. Drummond was especially vulnerable to bullshit, it was because he'd scored above average on the BRS, both in his responses to individual items as well as in his total score. Based on research showing an association between pseudo-profound bullshit receptivity and belief in conspiracy theories,[8] I also told the jury that Mr. Drummond's receptivity to bullshit made him more likely to buy into the convoluted fiction of sovereign citizen conspiracy theory. This placed him firmly in the middle of the disinformation food chain, where he was guilty of passing bullshit along to others but had also been duped by it himself.

Mr. Drummond's case highlights that, contrary to the familiar adage, "you can't bullshit a bullshitter," those who are purveyors of bullshit may just as often fall victim to it. This is exactly what University of Waterloo psychologist Shane Littrell and his colleagues found when they looked for associations between bullshit receptivity as measured by the BRS and scores on another questionnaire they developed called the Bullshitting Frequency Scale (BFS) that measures one's propensity to bullshit.[9] In their study, a specific type of bullshit called "persuasive bullshitting," that they defined as bullshitting intended to "impress, persuade, or fit in with others by exaggerating, embellishing, or otherwise stretching the truth about one's knowledge, ideas, attitudes, skills, or competence," was correlated with pseudo-profound bullshit receptivity. In other words, persuasive bullshitters both rely on bullshit to impress others, but are also more likely to be impressed when other people bullshit them. As I suggested in Chapter 5, and as I explained to Mr. Drummond's jury, this means that mesopredators within the disinformation food chain can be understood as having a kind of "epistemic naïveté" since they're often spreading bullshit—along with disinformation more generally—unintentionally and even unknowingly. To make matters worse, according to other research by Littrell and his colleagues, persuasive bullshitters often fall victim to a "bullshit blind spot"—like the "bias blind spot" that I mentioned in Chapter 3—whereby they're not merely oblivious to the fact that they're spreading bullshit, they're also falsely overconfident in their ability to detect it.[10] This supports what I suggested earlier—that bullshitting is subject to a Dunning-Kruger-like effect whereby those who are most sure that they're immune to bullshit are often those who are most likely to be taken in by it.

As a standalone trait, pseudo-profound bullshit receptivity can be summarized as the tendency to be impressed by words that might seem profound at first glance but are devoid of meaning on more careful examination. Not surprisingly then, Pennycook as well as other researchers have found that this kind of gullibility is associated with an "intuitive cognitive style."[4,11] Echoing the kind of "faith in intuition" that I mentioned in Chapters 1 and 6, "fast mode" thinking that I discussed in Chapter 2, and Stephen Colbert's "truthiness" that I described in Chapter 4, an intuitive cognitive style involves adopting beliefs based on how they sound or make us feel. Such prioritization of intuition above reasoning or evidence stands in contrast to analytical thinking, which is inversely correlated with bullshit receptivity and, as I noted in Chapter 4, has been found to be protective against belief in fake news and other kinds of misinformation. Indeed, Pennycook has reported evidence that bullshit receptivity is not only associated with belief in conspiracy theories, but also with the tendency to perceive false or fake news as accurate and a willingness to share it on social media.[12] Other research has also found a link between bullshit receptivity and belief in the paranormal, alternative medicine, and pseudoscience.[4,11,13]

Now that we've added bullshit receptivity to the list of other cognitive pitfalls that I've covered in this book thus far including unwarranted overconfidence, faith in intuition, confirmation bias, motivated reasoning, epistemic mistrust, and exposure to the barrage of both misinformation and disinformation that's out there in the world, we can start to appreciate how they are all interconnected within a maze of pathways that converge on false belief. Throughout the rest of this chapter, I'll explore how bullshit receptivity is exploited to manufacture trust in unreliable sources of information within several different arenas where bullshit is prevalent.

Scientific Bullshit and Pseudoscience

Deepak Chopra is, by almost every standard of public opinion, an epitome of success. A physician who completed his internal medicine residency in the United States after immigrating from India in the 1970s, he transformed himself from a practicing endocrinologist into one of the most recognizable faces of Transcendental Meditation and the New Age Movement thanks to promotion by Oprah Winfrey and notoriety as a friend and spiritual advisor to celebrities like Michael Jackson and Elizabeth Taylor. Touting meditation, yoga, and Ayurvedic medicine as keys to healthy living, he founded the Chopra Center for Wellbeing in the 1990s, established the nonprofit Chopra Foundation in 2006, and consolidated his ventures under the international brand Chopra Global in 2019. With an estimated personal net worth of $170 million, he has trademarked his name. His marriage has lasted more than 50 years, with two children and three grandchildren to show for it. To top it off, he has claimed to have never been sick a day in his life.[14]

Although part of Chopra's fortune has come from hawking herbal supplements with alleged anti-aging effects in the vein of Alex Jones, Kelly Brogan, and Joseph Mercola, he has been able to monetize his words above snake oil remedies in a way the others haven't mastered. With more than 90 authored works of fiction and non-fiction including 21 *New York Times* best-sellers to date, Chopra commands speaking fees of up to $35,000 per lecture and offers week-long retreats led by "Chopra master educators" costing attendees about $1,000 per day. At first glance, the broader appeal of his messaging seems to lie in steering clear of conspiracy theories or the wholesale bashing of science and Western medicine—he maintains professional membership in medical organizations like the American College of Physicians, holds academic appointments at Columbia Business School and the University of California at San Diego, and is pro-vaccine—in favor of bridging traditional Eastern and Western approaches to holistic medicine and health. A 2008 *Time* magazine feature dubbed Chopra a "New Age Supersage" and "man of science with the soul of a mystic" who found a formula for success in "packaging Eastern mystique in credible Western garb."[15] Indeed, the ad copy for Chopra Global boasts that his wellness empire

> is uniquely positioned at the intersection of science and spirituality and its signature programs are clinically-proven to impact numerous parameters of health for improved physical, mental and spiritual well-being. Informed by wisdom traditions and backed by cutting edge science, the Company encompasses a variety of products and services; from digital meditation apps to online masterclasses to live events and personalized retreats at the Chopra Center.[16]

No doubt, Chopra's enormous popularity can also be attributed to his promise of achieving the kind of superlatives that always seem to lie beyond our grasp—"stress-free living," "perfect health," "quantum healing," unleashing our "infinite potential" as a "metahuman," and "life after death." Just the kind of claims that, as I mentioned in the Preface of this book, have all the makings of a bestseller. And yet, a closer look at Chopra's words reveals that he's selling more than just the key to attaining the unattainable. Take this sentence from his most best-selling work, *The Seven Spiritual Laws of Success*:

> the physical universe is nothing other than the Self curving back within Itself to experience Itself as spirit, mind and physical matter.[17]

Or these two tweets from 2014:

> Attention and inattention are the mechanics of manifestation.
> Nature is a self-regulating ecosystem of awareneness.[18]

Or this explanation of "quantum healing," the subject of his 1989 book of the same name:

> Our bodies ultimately are fields of information, intelligence and energy. Quantum healing involves a shift in the fields of energy information, so as to bring about a correction in an idea that has gone wrong.[19]

And finally, this claim from a 2013 interview in *The Atlantic*:

> There are biofields—every part of our body, every cell of our body, has a magnetic field that it transmits. Our biofields are going to interfere with each other ... now we can biologically measure that. [You can] correlate states of consciousness with states of biology using mathematical algorithms and correlate that with crime, with hospital admissions, with traffic accidents, with social unrest, with quality of leadership.[20]

That millions of followers regard these words—which are barely distinguishable from the items included in Pennycook's BRS—as the deep thoughts of a New Age Supersage is testament to the real secret of Chopra's success. Word by word, lecture by lecture, and book by book, he's built a megamillion-dollar empire by selling the world on little more than pseudo-profound bullshit.

The widespread reverence that Chopra garners provides a sense of just how many of us have at least some degree of bullshit receptivity. Several years before the term "pseudo-profound bullshit" was coined, social and cognitive scientist Dan Sperber described our tendency to place blind trust in those who use obscure language to inspire awe as the "guru effect," noting that "all too often, [we] judge profound what [we] have failed to grasp."[21] Pennycook's subsequent research on pseudo-profound bullshit receptivity has since clarified that some are more vulnerable to this effect than others. Indeed, despite his mass appeal, Chopra has earned his share of detractors. A 2017 article in *Quartz* magazine noted that Chopra is both "beloved and despised" while being regarded as a "spiritual leader," "guru," and "prophet" by some and a "charlatan," "pseudoscientist," and "spouter of bullshit" by others.[18] Like Pennycook, Chopra was also awarded a 1998 Ig Nobel Prize for "interpretation of quantum physics as it applies to life, liberty, and the pursuit of economic happiness."[7] Of course, Chopra frames such criticism in an altogether different light—although some of the most vocal refutations of his dubious claims about quantum physics and evolution have come from renowned scientists like physicist Brian Cox and biologists Richard Dawkins and Jerry Coyne, Chopra dismisses those who accuse him of bullshit as "professional skeptics" who are "frozen in an obsolete worldview."[18]

* * *

When I was a medical student, I was assisting with the care of a patient with schizophrenia who was convinced that he'd figured out a new way of generating energy, similar to cold fusion, through the interaction of fire and ice. Based on his use of terms like "nuclear catalysis," "exothermic reaction," "deuterium," and "phase transition" and the fact that he was majoring in physics at a local community college, I had a moment when I wondered whether he'd actually stumbled on some kind of scientific breakthrough, providing evidence to support the romantic myth that people who suffer from psychotic disorders like schizophrenia may actually be covert geniuses. But when I looked over the pages of mathematical equations he'd sketched out with pencil and paper, it didn't seem to make any sense. Still, since I was several years removed from freshman physics at MIT at that point, I wasn't confident whether it might have been my fault for lack of understanding. So, I double-checked with my MIT friends who'd gone on to complete graduate degrees in physics and they agreed. It was nonsense. I have to admit, I was a little disappointed.

Social psychologist Anthony Evans and his colleagues at Tilburg University in the Netherlands argue that "scientific bullshit" deserves to be separated from pseudo-profound bullshit as a distinct variant. The former, they say, is "a form of communication that relies on obtuse scientific jargon to convey a false sense of importance or significance" that's intended to "sound true [rather than] profound."[22] To measure receptivity to scientific bullshit, they developed the Scientific Bullshit Receptivity Scale (SBRS) that includes 10 items "created by taking existing physical laws and changing their central words with randomly selected words from a physics glossary" such as:

- Energy can deteriorate based on closed-circuit alliterations of an afocal system.
- There are no transverse waves when the total magnetic sublimation through a stiff photon is equal to its scattered matrix.
- The solubility induced in an electromagnetic field is proportional to the damped vibration of the binary infrasound it encloses.

Evans's research found that scores on the SBRS varied across individuals, were significantly correlated with scores on Pennycook's BRS, and that scores on both the BRS and the SBRS were associated with faith in intuition as a cognitive style. Somewhat surprisingly though, it also found that both pseudo-profound bullshit receptivity and scientific bullshit receptivity were associated with "belief in science," defined as belief in "the value of science and its superiority as a source of knowledge." Another study by psychologists at the University of Zagreb likewise found that trust in science was a significant predictor of belief in COVID-19 conspiracy theories based on antiscience rhetoric.[23]

These counterintuitive results reveal that instead of being protective against bullshit, as we might expect, faith in science can actually make us more vulnerable to it. Taken together with the aforementioned finding that pseudo-profound bullshit receptivity is associated with belief in pseudoscience, a clearer picture emerges of why Chopra's words—not unlike the theories of my patient who was convinced he'd made a breakthrough in physics—might sound so convincingly meaningful at first glance. Pseudoscience has been defined as "a form of imitation or fakery" that "exhibits the superficial trappings of science," but offers "epistemic fool's gold."[24] In other words, it's a kind of bait-and-switch whereby bullshit is dressed up in scientific jargon to give it a legitimacy that it doesn't deserve. When Chopra "packages Eastern mystique in credible Western garb," he's cloaking Eastern mysticism in terms like "quantum," "photons," and "consciousness" to make it sound "sciencey."

This approach isn't novel or unique to Chopra. As a self-described practitioner of "alternative," "complementary," or "integrative" medicine, he's part of a much larger movement that attempts to capitalize on the shortcomings of science and Western medicine that I acknowledged in Chapter 1 by placing pseudoscientific claims on equal footing with, or above, those of science. As I discussed in Chapter 5, this is the same strategy and business model that's made Gwenyth Paltrow's Goop a multimillion-dollar empire, except that Paltrow often falls back on portraying herself as a prosumer who's "just asking questions" instead of marketing herself as a guru the way Chopra does. But either way, substituting "sciencey" words for legitimate scientific concepts is a tactic taken straight out of the disinformation playbook that substitutes "truthiness" for truth. Claims about the validity of "alternative medicine" are much the same as claims that "alternative facts" are interchangeable with actual facts. But, as the meme goes, "if alternative medicine actually worked, we'd just call it medicine."

* * *

To be clear, research has found that some forms of alternative medicine like mindfulness meditation or yoga can have positive health effects. But other interventions falling under the broad umbrella of alternative medicine have crumbled under more careful scientific scrutiny. For example, when Chopra talks about "biofields" in the context of "quantum healing," he's suggesting that the weak electromagnetic energy—mostly in the form of infrared radiation—that is indeed emitted from our bodies can somehow be harnessed or modified for the purposes of healing. Some alternative medicine practitioners claim that the existence of these fields provides scientific validation of the "life forces" such as *prana* and *chi* that are central to Eastern medical traditions. These "human energy fields" have, in turn, formed the basis for an alternative medicine treatment modality called Therapeutic Touch (TT) that, contrary to its name, doesn't actually involve any touching. According to one of its pioneers, the late New York University nursing professor Dolores Krieger, it entails having practitioners use their hands, held over a patient's body, to sense imbalances in their "nonphysical human energy" based on the detection of "sensory cues" in the form of "vague hunches, passing impressions, flights of fancy, or, in precious moments, true insights or intuitions."[25] Once so detected, the energy fields are "rebalanced," "repatterned," or "remodulated" by "removing congestion, replenishing depleted areas, and smoothing out ill-flowing areas."[26] At the height of its popularity through the 1990s, TT was endorsed by the likes of the National League for Nursing and the American Nurses' Association and was reportedly widely taught in colleges and nursing schools to tens of thousands of practitioners.

In 1996, however, 9-year-old Emily Rosa was brainstorming experiments for her fourth-grade science fair and decided to test the efficacy of TT after learning about it from a video.[27] She devised a simple test to see if TT practitioners could really detect human energy fields. Using herself as the would-be patient, she sat across from experienced TT practitioners on the other side of a cardboard barrier so that they couldn't see each other. The practitioners placed their arms through holes in the barrier with their palms facing upward and Rosa, based on a flip of a coin, placed her hand above either the right or left hand of the practitioner and asked them which of their hands was closest to hers. After all, if TT practitioners could really detect her energy fields as they claimed, they should have been able detect it under the blinded conditions of her experiment. But across 21 TT practitioners and 280 attempts, they were only able to answer correctly 44% of the time—that is, no better than chance.

With her mother as lead author, Rosa got her study published in *JAMA* and enjoyed a flurry of national media attention as a result, with recognition by the Guinness Book of World Records as the youngest person to ever be published in a medical journal at the time. When Kreiger was named the winner of the Ig Nobel Prize in Science Education in 1998 (the same year Chopra got his for physics) for "demonstrating the merits of therapeutic touch, a method by which nurses manipulate the energy fields of ailing patients by carefully avoiding physical contact with those patients"[7] and didn't accept the award, Rosa was invited to accept it in her stead and deliver the ceremony's keynote address. Rosa, for her part, downplayed her accomplishment in an article published in *Jr. Skeptic* magazine, writing, "I only did basic (also cheap and fairly easy) research on TT, something nurses should have thought about doing a long time before I was even born."[27] She went on to write,

People who sell pseudoscience like TT usually don't think it's important to test their theories, and adult scientists often don't have time for it. We kids can learn science by testing pseudoscience, and sometimes I think it's not as easy to fool kids. If you look closely at pseudoscience, I bet you, too, can figure out an easy test. You might catch adults doing silly, foolish, or even criminal things—or find something real.

Although Chopra acknowledges that TT and other forms of "energy healing" based on biofields are "highly controversial" in Western medicine, he has nevertheless promoted it based on claims that it's supported by "scientific studies ... using accepted scientific methods."[28] Curiously though, in an article he wrote defending "biofield healing," he cited only a single study that actually found no difference between a form of TT called "energy chelation" and "mock healing" as a placebo control.[29] Rosa's JAMA paper included a much more thorough review of the clinical trial data on TT and noted that "the methods, credibility, and significance of these studies have been seriously questioned."[26] That matches the conclusion of more recent reviews of TT that have found little in the way of convincing evidence to support that it offers anything beyond a placebo effect.[25,30]

While it could be argued that the placebo effects associated with alternative medicine represent safe and effective clinical interventions worthy of additional research, the potential placebo effects of TT's handwaving do nothing to validate the premise that biofields can be manipulated as a therapeutic modality. And worse, as I mentioned in Chapter 5, there are dangers in legitimizing such interventions. First, when people seek out alternative interventions of dubious efficacy against medical illness, it often comes at the expense of bypassing more effective medical treatment. Second, as I briefly mentioned in Chapter 5, "alternative New Age spirituality" has found a concerning bedfellow in conspiracy theory belief over the past several years, an overlap that has been labeled "conspirituality."[31]

A recent study by psychologists at the University of Amsterdam found that belief in COVID-19 conspiracy theories wasn't predicted by psychological needs like the "three C's" or need for uniqueness that I reviewed in the previous chapter but by lower levels of analytical thinking and higher levels of spirituality as measured by endorsement of statements like "I meditate to gain access to my inner spirit."[32] Indeed, during the pandemic, many within the "wellness" movement, including those dedicated to spirituality, yoga, and alternative medicine, found themselves drawn into QAnon, gaining initial entry through the soft-core New Age narratives of so-called "pastel QAnon" before moving into harder core conspiracy theories through a shared attraction to themes of mistrust in Western medicine and the institution of science along with receptivity to bullshit.[33] Although those interested in spirituality and wellness are focused on attaining a kind of holistic health to which many of us should aspire, too often, when pseudoscience and bullshit receptivity meet in this domain, it ends up being on the dark side of false belief. False beliefs and bullshit aren't the solution for what ails us, as Chopra seems to suggest. They're the cause.

Unlike many of the other cognitive quirks underlying false belief covered thus far, Pennycook and others have found that bullshit receptivity is consistently associated with lower cognitive ability and intelligence as measured by performance on a variety of standardized tests.[4,8,11,13] As a result, media coverage of the study that earned Pennycook an Ig Nobel Prize ran headlines like:

> Do You Love "Wise-Sounding" Quotes? Surprise! You're Probably Dumb.[34]
>
> People Who Post Inspirational Facebook Quotes Are Morons, According to Science[35]

Such headlines are misguided interpretations of what "lower cognitive ability" really means however. In reality, cognitive ability is much less an immutable trait than a changeable state. The kind of analytical thinking that serves as an antidote to bullshit receptivity and belief in misinformation more generally is a skill, like "slow mode" thinking, that's achievable by all of us, as I explain more thoroughly in Chapter 10. In the meantime, although the "bullshit asymmetry principle"—also called Brandolini's Law after its originator Alberto Brandolini—claims that "the amount of energy needed to refute bullshit is an order of magnitude bigger than [what's needed] to produce it," Emily Rosa shows us that it can be accomplished nonetheless. If a 9-year-old girl can prevail over bullshit, so can we all.

Postmodernist Bullshit in the Ivory Tower

In 1996, a paper written by New York University physicist Alan Sokal entitled "Transgressing the Boundaries: Toward a Transformative Hermeneutics of Quantum Gravity" was published in the academic journal of cultural studies *Social Text*. The article purported to be a scientist's critique of science, drawing from the philosophical and literary traditions of postmodernism, poststructuralism, feminist studies, and critical theory to argue the social relativism of physical science—that is, against what Sokal's paper called the "façade" that there's any such thing as physical reality or objective truth. The opening paragraph began,

> There are many natural scientists, and especially physicists, who ... cling to the dogma imposed by the long post-Enlightenment hegemony over the Western intellectual outlook, which can be summarized briefly as follows: that there exists an external world, whose properties are independent of any individual human being and indeed of humanity as a whole; that these properties are encoded in "eternal" physical laws; and that human beings can obtain reliable, albeit imperfect and tentative, knowledge of these laws by hewing to the "objective" procedures and epistemological strictures prescribed by the (so-called) scientific method.[36]

Densely footnoted and referenced, it went on to propose that Einstein's theory of relativity provided evidence to support a wide range of academic theories from the humanities with profound implications spanning politics, philosophy, and literary criticism. For example, it claimed that the deconstructionist philosopher Jacques Derrida's passing comments referring to the "Einsteinian constant" and "variability" were reflective of a physics in which

> any space-time point, if it exists at all, can be transformed into any other. In this way the infinite-dimensional invariance group erodes the distinction between observer and observed; the π of Euclid and the G of Newton, formerly thought to be constant and universal, are now perceived in their ineluctable historicity; and the putative observer becomes fatally de-centered, disconnected from any epistemic link to a space-time point that can no longer be defined by geometry alone.[36]

It similarly claimed that the psychoanalyst Jacques Lacan's words were supported by the mathematics of typology and that commentary by the feminist Luce Irigaray conformed with string theory. Finally, it concluded with these sentences speculating on the future of physics and mathematics:

> [A] liberatory science cannot be complete without a profound revision of the canon of mathematics. As yet no such emancipatory mathematics exists, and we can only speculate upon its eventual content. We can see hints of it in the multidimensional and nonlinear logic of fuzzy systems theory; but this approach is still heavily marked by its origins in the crisis of late-capitalist production relations.
>
> Catastrophe theory, with its dialectical emphases on smoothness/discontinuity and metamorphosis/unfolding, will indubitably play a major role in the future mathematics; but much theoretical work remains to be done before this approach can become a concrete tool of progressive political praxis. Finally, chaos theory—which provides our deepest insights into the ubiquitous yet mysterious phenomenon of nonlinearity—will be central to all future mathematics.[36]

Sokal's paper had a seismic impact across academia and garnered international media coverage, but not in the way the editors of *Social Text* anticipated, because it turned out that the whole thing was a hoax. In a companion piece that came out just a few weeks later in the journal *Lingua Franca*, Sokal described the "fundamental silliness" of his *Social Text* article, noting that his claims that "quantum physics is profoundly consonant with 'postmodern epistemology'" or that "postmodern science [had] abolished the concept of objective reality" were "liberally salted with nonsense" and contained "only citations of authority, plays on words, strained analogies, and bald assertions" that were devoid of "anything resembling a logical sequence of thought."[37] He explained that while his method was "satirical" parody, his motivation was "utterly serious" as a protest against what he saw as

> the proliferation, not just of nonsense and sloppy thinking per se, but of a particular kind of nonsense and sloppy thinking; one that denies the existence of

objective realities ... [and] contemporary academic theorizing [that] consists precisely of attempts to blur these obvious truths—the utter absurdity of it all being concealed through obscure and pretentious language.

Bristling at "radical relativism"—the postmodern premise that reality is but a social construct—Sokal added that "we [can't] combat false ideas in history, sociology, economics and politics if we reject the notions of truth and falsity."

Just what exactly was Sokal up to? In short, he was using scientific and persuasive bullshit to protest against what he saw as pervasive bullshit within the cultural studies departments of academia. He was trying to show that if the editors of *Social Text* couldn't distinguish his bullshit from meaningful insights, it meant that everything else published in the journal and much of what was coming out of cultural studies at the time might very well be bullshit, too.

* * *

The Sokal Affair illustrates several important aspects of bullshit. For one thing, it reinforces the observation that bullshit is all around us. It's not only found in the rhetoric of Deepak Chopra's and Gwyneth Paltrow's modern New Age pseudoscience—it's been a core feature of the much longer-standing academic traditions of postmodernism and poststructuralism that are now nearly a century old. These intellectual movements began as a kind of revolt against both the Age of Enlightenment, during which time science and reason claimed ascendancy, as well as against modernism, when industrialization and technological advancements facilitated the advancement of civilization into two world wars. Postmodernism has been defined as

> largely a reaction to the assumed certainty of scientific, or objective, efforts to explain reality. In essence, it stems from a recognition that reality is not simply mirrored in human understanding of it, but rather, is constructed as the mind tries to understand its own particular and personal reality. For this reason, postmodernism is highly skeptical of explanations which claim to be valid for all groups, cultures, traditions, or races, and instead focuses on the relative truths of each person. In the postmodern understanding, interpretation is everything; reality only comes into being through our interpretations of what the world means to us individually. Postmodernism relies on concrete experience over abstract principles, knowing always that the outcome of one's own experience will necessarily be fallible and relative, rather than certain and universal. Postmodernism is "post" because it denies the existence of any ultimate principles, and it lacks the optimism of there being a scientific, philosophical, or religious truth which will explain everything for everybody.[38]

By the 1990s, when Sokal wrote his paper, science was on the rise in academia while postmodernism was in decline, but its embers were still being stoked within certain disciplines. As the philosopher Daniel Dennett noted,

Postmodernism, the school of "thought" that proclaimed "There are no truths, only interpretations" has largely played itself out in absurdity, but it has left behind a generation of academics in the humanities disabled by their distrust of the very idea of truth and their disrespect for evidence, settling for "conversations" in which nobody is wrong and nothing can be confirmed, only asserted with whatever style you can muster.[39]

After reading biologist Paul Gross and mathematician Normal Levitt's 1994 book *Higher Superstition: The Academic Left and Its Quarrels with Science* that argued that postmodernism's critiques of science included claims that its discoveries were merely social constructs or that objective truth didn't exist, Sokal became concerned and perhaps even annoyed. As he wrote in his *Lingua Franca* piece, "anyone who believes that the laws of physics are mere social conventions is invited to try transgressing those conventions from the windows of my [twenty-first floor] apartment."[37] This, you might recall, is much the same sentiment I expressed in the Preface of this book when I noted that categorical distinctions between colors, rather than being arbitrary or relativistic, become all too real when deciding what to do at a traffic light.

The Sokal Affair also reinforces what I suggested earlier—that bullshit receptivity isn't just a cognitive trait of "morons" or "idiots" as is often claimed. The four editors who reviewed and published Sokal's paper were highly educated academics with advanced degrees in their fields, so that it wasn't a deficit of cognitive ability or intelligence that made them fall for bullshit. While we could chalk it up to a real-life example of the "bullshit blind spot" whereby bullshitters got taken in by bullshit themselves, understanding why the editors were fooled also requires that we consider both the content of the paper as well as the context in which it was reviewed. With Sokal's treatise littered with as much scientific bullshit as postmodernist bullshit, it's likely that the *Social Text* editors were out of their depth in reviewing Sokal's scientific claims and accepted them at face value given his position as a university physicist. As I described in Chapter 5, decisions to trust the information that we hear from others is based on an assessment of "source credibility." It has been shown that source credibility has a significant effect on bullshit receptivity such that our failure to recognize bullshit per the guru effect increases when its alleged source is recognized as a famous or otherwise noteworthy person.[40] Sokal's academic credentials no doubt carried a lot of weight with the editors of *Social Text*.

Ironically, given the title of the journal in which Sokal's paper appeared, the social context of his submission was probably just as relevant to its publication as its content and source, if not more so. As Bruce Robbins, a *Social Text* editor at the time of Sokal's paper who wasn't one of its reviewers, explained in 2017, the article was well-received because Sokal—a scientist—had staked out a critical postmodern position against science that amounted to switching sides in an ongoing cultural war.[41] It was featured as the concluding piece within a special issue of *Social Text* called "The Science Wars" that addressed the long-standing tension between science and postmodernism. According to University of Santa Cruz history professor Barbara Epstein, the editors who read and accepted Sokal's piece weren't merely blinded due to bullshit receptivity, they were blinded because it was such a "coup" that a scientist had come to postmodernism's defense that they "didn't bother to read [the] article carefully."[41] This account suggests that receptivity to bullshit can occur through both a fast intuitive mode of thinking as well as a slower mode of thinking that includes motivated reasoning. As it turns out, this is exactly what inspired Sokal's hoax—he was trying to prove that some academics are so desperate to see their ideas validated that it doesn't matter if that validation comes in the form of bullshit. Just so, the 1996 Ig Nobel Prize for Literature didn't go to Sokal but to the editors of *Social Text* "for eagerly publishing research that they could not understand, that the author said was meaningless, and which claimed that reality does not exist."[7]

All of this leaves us with the real conundrum of bullshit—how to distinguish it from technical jargon and advanced concepts that lie beyond our comprehension but are in fact insightful and profound. The answer should be the same regardless of who we are—when we aren't sure what's true, we can run it by those who have more expertise on the subject than we do and are in a good position to offer an unbiased critique. When I was skeptical of my patient's theories and mathematical equations that he believed formed the basis for a revolutionary new way to produce energy, I checked with my physicist friends who knew far more about the subject than I did and I listened to them. When experts in their fields like Brian Cox, Richard Dawkins, and Jerry Coyne tell us that Deepak Chopra's musings about quantum physics and evolution amount to scientific bullshit, we should listen to them, too. What happened when Sokal's paper was read by the editors of *Social Text* was something different. The kind of objective and blinded peer review process that I described in Chapter 4 that's supposed to take place when an academic journal reviews a paper never happened. Instead, the editors, who were cited in the paper in flattering ways and had skin in the game, just read it over themselves. That left them wide open to bullshit that they then passed on to a larger audience.

Evasive Bullshitting and Politics

Although Sokal is a self-described "leftist and feminist" who at one time taught mathematics in Nicaragua to support the Sandinista National Liberation Front, he made clear in his *Lingua Franca* reveal that the "epistemic relativism" and "obscurantism" of 1990s academia was largely "emanating from the self-proclaimed Left."[36] Most of his critical attention since has likewise remained focused on postmodern thinking within liberal academic circles because, as he sees it, the "leftist and feminist" ideals that he believes in are supported by "evidence and logic" that can be explained with clear language and without the need for "self-indulgent nonsense."

Over the quarter century since the Sokal Affair, Sokal has continued to be a critic of bullshit in academia. But in 2006, he posed the rhetorical question, "does it really matter?," asking whether postmodernist argumentation is just harmless intellectual theory.[42] He answered by drawing a parallel between postmodernism and pseudoscience—the two forms of bullshit I've addressed in this chapter thus far—noting that they're often linked together as "fellow travelers," with pseudoscientists using postmodernist arguments to defend their claims and postmodernists supporting them. He then concluded that the connection between them is "most dangerous when they are conjoined to political movements." But when he connected the dots between postmodernism, pseudoscience, and politics, he wasn't pointing his finger at liberal academia this time—he was citing historical examples of those entities converging within Nazi Germany and India's Hindu Nationalism. So, while Sokal recognized that there was something rotten in liberal postmodernism that needed fixing, he made clear that the danger of such bullshit in politics extended to conservative movements as well.

And yet, what Sokal doesn't seem to have ever intended, much less anticipated, was that his initial attempt to get his university humanities colleagues' house in order might have such far-reaching ripple effects within post-truth politics in the world today. In the early wake of the Sokal Affair, conservative commentators like George Will and Rush Limbaugh had a field day, seizing upon its object lesson to rail against the snooty faux-intellectualism and highfaluting pretension of ivory tower liberalism. As if sparked by Sokal's gotcha exposé, that conservative-turned-populist perspective has spread into something more pervasive and dangerous today, with intellectuals, elites, experts, and scientists lumped together and discounted en masse—seemingly based on guilt by academic or political association—well beyond the confines of university humanities departments. Ironically then, Sokal's attempt to defend science by exposing postmodernist bullshit may have helped pave the path to the new variant of postmodernism that plagues us now—a post-truth world in which anything, even scientific evidence, is routinely opposed by "alternative facts."

* * *

While today's post-truth politics is postmodern in spirit, it doesn't hide behind postmodernism's inscrutable rhetoric. On the contrary, political speechmaking these days often makes a point of avoiding that pitfall and steering well clear of it. When the likes of Boris Johnson, Donald Trump, and Hillary Clinton are called out as master bullshit artists,[43] no one's accusing them of spouting pseudo-profound bullshit, just as no one is mistaking their words for anything particularly profound. Any claim that bullshit is pervasive in post-truth politics today is therefore either conflating bullshitting with lying or talking about a different kind of bullshit.

Shane Littrell, who revealed that bullshitters may be especially vulnerable to bullshit themselves, argues that bullshit can be most meaningfully differentiated based not on its content, like "pseudo-profound" or "scientific," but by its intent. In his BFS scale that quantifies one's propensity to bullshit, the kind of "persuasive bullshitting" that I discussed earlier is distinguished from "evasive bullshitting," defined as "strategic evasiveness or bluffing" that acts as a kind of cover for ignorance or to hide the truth.[9] Although politicians use persuasive bullshitting to prop themselves up often enough, evasive bullshit that's used to avoid answering questions honestly or maintain a façade of knowledge is at least as common in politics. We see it all the time when politicians answer questions in public, whether in an election debate or a press conference.

When Littrell and his colleagues compared persuasive and evasive bullshit in their study and looked for associations with other variables, they found several key differences. While persuasive bullshitting was positively associated with overconfidence and negatively associated with cognitive ability, evasive bullshitting demonstrated the reverse pattern.[9] Furthermore, unlike persuasive bullshitting, evasive bullshitting wasn't associated with receptivity to pseudo-profound or scientific bullshit—instead, evasive bullshitting was linked to a *lower* risk of falling for bullshit. If evasive bullshitters have greater cognitive ability and lack the bullshit blind spot that persuasive bullshitters often have, then it might be appropriate to think of them as less epistemically naïve and therefore more deceitful.

Although bullshitting of any type is defined as distinct from lying, its indifference to truth or falsity doesn't necessarily make it the lesser of the two evils. On the contrary, when Frankfurt wrote about bullshit back in 1986, he noted that "bullshit is a far greater enemy of the truth than lies can ever be."[2] Echoing what Sokal said about the danger of bullshit in the form of postmodernism and pseudoscience conjoining with political movements, it's been argued that, in politics, bullshitting is more harmful than lying because it's inherently ambiguous and lacking in substance to the point of being resistant to either refutation or fact-checking.[43] In other words, political leaders who have mastered the art of evasive bullshitting can't be held accountable. As Shakespeare told us, "words are but wind"—and that goes double for evasive bullshitting so that it often doesn't seem to matter what political leaders say anymore. Is that really the kind of low bar we want to set for our elected government representatives?

Calling Bullshit

As a medical student, I don't recall ever being taught that patients sometimes lie about their symptoms or make things up. But in psychiatric training, and especially in forensic psychiatry, it's well-understood that people sometimes "malinger" or feign symptoms of physical or mental illness for some tangible benefit, such as to obtain a disability income, gain admission to the hospital, or escape culpability for crime. Detecting malingering—that is, figuring out if someone is lying about their symptoms or not—is often challenging, especially when someone is making claims about psychiatric experiences that are subjective and internal like "hearing voices" or having panic attacks.[44] After all, unlike other medical specialists, psychiatrists rarely make diagnoses based on blood tests and brain scans—we have to rely on careful psychiatric interviewing. During that interview, patients with real experiences and symptoms can usually put them into words in a coherent and consistent fashion. By contrast, the typical "red flags" of malingering are vagueness, evasiveness, and inconsistency when answering questions about purported symptoms—the more details the interviewer asks about, the less is learned. In other words, when malingerers lie, they're defending their lie, or covering it up, with evasive bullshit.

Although malingering was at one time modeled as a form of mental illness, it's since been increasingly recognized as a potentially adaptive behavior that's often motivated by social circumstances and pressures. For example, with fewer and fewer inpatient beds available these days, hospital admission criteria have become increasingly stringent so that a patient who comes to the ER saying that they're merely "depressed" isn't as likely to be admitted as someone who says they're "suicidal" or "hearing voices." And so, if a patient really wants to be admitted—for example, to get off the street as respite from homelessness—they know they should say something like "I'm hearing voices telling me to kill myself" even if that's not really the case. Although physicians can often feel slighted or resentful when they're lied to in this way, my colleagues and I coined the term "iatrogenic malingering" in 2003 to highlight how we shouldn't blame patients for misrepresenting the truth when we're the ones—through healthcare rationing across hospital systems—who both incentivize and reinforce such behavior.[45] By the same token, if malingering is "iatrogenic"—that is, caused by an intervention—it means that the best way to treat it is to change the conditions that incentivize it. If patients feel like they need to lie to get the help they want, physicians and healthcare systems need to figure out how to get them the help they need without having to lie.

University of London professor of organizational behavior André Spicer has presented a similar explanation for why bullshit is so prevalent within certain "speech communities" like business organizations, calling bullshit a "form of social interaction which is accepted or even encouraged within a particular community."[46] That echoes what John Petrocelli told us at the beginning of this chapter about resorting to bullshit when there's a social pressure to provide an opinion and a social pass that allows one to get away with a nonsensical response. Like iatrogenic malingering then, if there's so much lying and bullshit around, it's because we incentivize it and allow it to happen.

Indeed, there may be no bigger incentive or allowance for lying and bullshit alike than in politics. From 2016 to 2019, when a Pew poll asked respondents if they had either a fair amount or a great deal of trust in various public figures, only 25–35% trusted elected officials, trailing by about 10% behind business leaders and the news media.[47] No doubt this is because we don't tend to hold politicians to a high standard of truth-telling—when we watch political debates these days, we've grown accustomed to hearing two different accounts of reality and have come to rely on post-debate fact-checking as a necessity. When *PoliFact* editor-in-chief Angie Drobnic Holan tallied the accuracy of statements made by US presidential candidates from 2007 to 2015, all of them had uttered falsehoods and half-truths.[48] Although she found that Democratic candidates tended to stick closer to the truth than Republicans overall, the most honest among them still only told the truth about 50% of the time. Although Ben Carson and Donald Trump topped the list as the most untruthful politicians, the lowest percentage of falsehoods was owned by Bill Clinton, a dubious distinction given that he was impeached for lying. Whether we're talking about persuasive bullshitting, evasive bullshitting, or lying, playing fast and loose with the truth is so common within the political "speech community" that few of us bat an eye when we hear it.

Another explanation for why bullshitting is so common in politics brings us back full circle to bullshit receptivity. Pennycook's research suggests that some of us are better at detecting bullshit than others due to the extent to which we rely on intuition versus analytical thinking—as a result, some can smell it from a mile away while others consume it without seeing it for what it is. Meanwhile, other research has looked to see whether bullshit receptivity might be linked to political ideology, with at least five studies to date finding an association with political conservatism.[21,49-52] One even found that bullshit receptivity was associated with favorable views of 2016 presidential candidates Ted Cruz, Marco Rubio, and Donald Trump, but not Hillary Clinton or Bernie Sanders.[50]

Still, there's reason to be circumspect about the claim that conservatives are more prone to bullshit receptivity than liberals. For one thing, bullshit receptivity has been consistently linked to conservatism on social issues but not economic conservatism. In addition, while a study published by researchers in Sweden—where there's far more political party diversity than there is here in the United States—did find support for an association between bullshit receptivity and conservatism, it also found that bullshit receptivity was actually highest among left-wing Green Party supporters who tended to also support "a number of unfounded ideas, such as astrology, alternative medicine, the moon landing conspiracy theory, and electric allergy."[51] Taken together with the Sokal Affair and the fact that being a fan of Deepak Chopra's is hardly a conservative stereotype, we would do well to avoid the easy trap of claiming that bullshit receptivity is the exclusive domain of any one political orientation. The relationship between bullshit receptivity and political ideology is more complicated than that.

Fortunately, Petrocelli devised a clever experiment to illuminate that complexity. He administered Pennycook's BRS to research subjects but randomly attributed the individual BRS items to either "Democratic leaders" like Barack Obama, Bill Clinton, Nancy Pelosi, and Oprah Winfrey or "Republican leaders" like George W. Bush, Paul Ryan, John McCain, or Rush Limbaugh.[52] Although there was an overall association between bullshit receptivity and conservatism, subjects were more susceptible to bullshit when the source was attributed to leaders of their own political party, whereas they were more likely to identify the same items as bullshit when they were attributed to leaders of their opposing party, regardless of whether the subjects self-identified as liberals or conservatives. As Petrocelli explained, the "statements allegedly made by leaders of one's political affiliation that are viewed as relatively profound are the very same statements made by leaders not of one's political affiliation that are viewed as bullshit." This conclusion mirrors the interaction between bullshit receptivity and source credibility that I mentioned earlier as well as the role of motivated reasoning in conspiracy theory belief that I described in the previous chapter—when it comes to political bullshit, bullshit receptivity is conditional depending less on what's said and more on who said it. Put another way, we incentivize and reinforce bullshit when it's coming from people we admire or those who seem to align with our beliefs and values. Whether it's Deepak Chopra making claims about quantum healing, Alan Sokal defending postmodernism in *Social Text*, or politicians trying to dodge questions by obfuscating, they're serving up bullshit because their audience eats it up.

***

As with the other cognitive pitfalls that make us vulnerable to false belief, bullshit receptivity isn't an "all or none" trait that exists only in certain individuals like our political or ideological opposites. But while we're all susceptible to bullshit to some degree, we all have it within ourselves to resist it if we try. How can we do that? For one thing, we can teach ourselves to be on the lookout for bullshit and to be honest about our own propensity to fall for it, keeping in mind that those who think they're immune are often the most vulnerable. Next, we can train ourselves to be more like Emily Rosa, using a bit of analytical thinking to sniff out bullshit and avoid being duped by it. As with findings about countering cognitive laziness with analytical thinking to help us become better at detecting "fake news," as I discussed in Chapter 4, Littrell and his colleagues recently found that engaging in "explanatory reflection" by pausing to ask ourselves *why* a statement sounds profound can reduce our receptivity to pseudo-profound bullshit, perhaps by not allowing ourselves to so easily pretend that we understand the inscrutable.[53] Unfortunately, this simple approach wasn't effective against scientific bullshit, and both pseudo-profound and scientific bullshit receptivity were greater when statements were falsely attributed to experts, consistent with previously mentioned findings about source credibility. This suggests that we need something more than just analytical thinking and explanatory reflection at the level of the individual to counter the effects of misplaced trust in gurus and other would-be experts. As I discuss further in Chapter 10, interventions at the societal level may be necessary to combat bullshit and the larger problem of misinformation as a contributor to false belief, starting with collectively pushing back against the postmodern claim that truth is relativistic, political, or otherwise endlessly open to interpretation or debate and instead embracing the antithesis of bullshit—the idea that truth exists and matters. Next, we can commit to removing the incentives for bullshit and demand that our political, academic, scientific, and spiritual leaders—regardless of whether they're on "our side" or not—make clearly stated claims backed up by evidence and hold them accountable when they don't stand up to fact-checking and peer review.

In 2017, University of Washington professors Carl Bergstrom and Jevin West, who define bullshit more broadly as "language, statistics, data graphics, and other forms of presentation intended to persuade by impressing and overwhelming a reader or listener, with a blatant disregard for truth and logical coherence," started a college course called "Calling Bullshit" that teaches students "data reasoning in a digital world."[54] Scientists by training, Bergstrom and West are primarily concerned about what they call "new-school bullshit" that "uses the language of math and science" to give a "veneer of legitimacy by glossing them with numbers, figures, statistics, and data graphics."[55] Here's how they explain their mission to help people see through this new type of scientific bullshit and its use to advance political agendas:

> We have a civic motivation.... It's not a matter of left- or right-wing ideology; both sides of the aisle have proven themselves facile at creating and spreading bullshit. Rather (and at the risk of grandiose language) adequate bullshit detection strikes us as essential to the survival of liberal democracy. Democracy has always relied on a critically-thinking electorate, but never has this been more important than in the current age of false news and international interference in the electoral process via propaganda disseminated over social media.[54]

"Calling Bullshit" focuses on one type of bullshit and is offered in only one college course at one university at the moment, but Bergstrom and West launched a website with an open course syllabus and published a book with the same name. So perhaps there's hope—the blueprint to lead us away from bullshit is laid out there, ready to be adapted to a larger and broader scale. We just have to choose to follow it instead of imagining that it's only something that other people need.

8
Divided States

> The purpose of politics is to enable people who disagree to live together.
> —Bob Waterman

A Marriage Strained

When Stephanie reached out to me, she was at her wit's end about her marriage. She was one of the many readers I mentioned near the end of Chapter 6 who'd contacted me for advice after I'd written a three-part series about how to help loved ones climb out of the QAnon rabbit hole.[1] Stephanie and her husband Nick had been together for about eight years and throughout the first half of their marriage their political differences—he's a conservative while she describes herself as "middle of the road"—made for interesting conversation and the occasional spirited debate. She felt they had a strong relationship that was able to tolerate whatever disagreements had come up through the years, but things took a turn when Donald Trump was running for president. Stephanie was put off by Trump's "grab 'em by the pussy" comments while Nick was getting increasingly worked up over Hillary Clinton's emails or her involvement with Jeffrey Epstein and the Democrats' alleged child sex trafficking ring. Together they were finding it harder to laugh it off when Stephanie told Nick he was crazy for believing Pizzagate or when she pointed out things like Trump's own ties to Epstein.

When Trump won the 2016 election, Stephanie expected Nick to be happier—he'd voted for Trump and she hadn't—but instead he just seemed to be getting more upset. And not just about political issues like the need to "build a wall" to halt immigration or the Mueller investigation "witch hunt," but at Stephanie in a way that he never had been before when they didn't see eye to eye. Then, in 2020, during the spring of the COVID-19 pandemic, they both started working from home and found themselves increasingly at odds. Nick was spending more time on YouTube viewing content by Ben Shapiro, PragerU, and Stefan Molyneux and he didn't like that Stephanie wouldn't watch along with him and was sticking to mainstream news instead. By the summer, when protests and rioting spread across the country in the wake of George Floyd's death, Nick was denouncing the Black Lives Matter movement as a terrorist organization and saying that Trump should send in the National Guard to "take out" Antifa. When Stephanie expressed her sympathy for the protesters, Nick started calling her a "socialist" and accusing her of being a "sheep" who was "blue pilled." During one argument, Stephanie told Nick that she felt like he was increasingly treating her like "the enemy" whereas he said that he just wanted her to "wake up" to the truth and be on his side. Finally, in the lead-up to the 2020 election, with Stephanie planning to vote for Joe Biden and Nick talking more and more about QAnon, they reached an impasse. Both of them felt like they were stuck between a rock and a hard place where the survival of their marriage seemed to depend on one of them changing their political beliefs.

When Stephanie wrote to me, she wanted to know what she could do to save her marriage and get her husband back. But other than referring her back to what I'd written on the subject and pointing her to a few other resources, there wasn't much I could say. For one thing, Stephanie and Nick weren't my patients, and I don't as a rule offer personal advice to readers. But, even if I did, Stephanie's dilemma brought to mind the joke that asks the question, "How many psychiatrists does it take to change a lightbulb?" and answers, "Just one, but the lightbulb has to really want to change." Stephanie wanted Nick to climb out of the rabbit hole where she believed he was being egged on by misinformation, but Nick didn't want to come out—he felt he'd found something of vital importance and wanted her to jump down there with him. Neither wanted to change their political beliefs; they wanted the other person to change theirs. I had no idea how Stephanie and Nick's marriage was going to turn out, but, as a starting point, that didn't leave a lot of room to reconcile their ideological differences.

Identity Politics

Many other accounts of families like Stephanie's being split apart by QAnon have been covered in the media over the past several years along with advice—some of it solicited from me—about how to navigate our interactions with relatives in a new era of political and ideological division.[2] But the division affecting families and relationships across the country is less a story about QAnon or conspiracy theories per se than it is one of the polarization of mainstream politics and a growing inability for ideological opposites to find common ground and work to preserve a communal identity.

Within several previous chapters, I've mentioned that false beliefs and politics often go hand in hand due to confirmation bias and motivated reasoning according to the echo chambers in which we find ourselves. While I've acknowledged that research has occasionally found that phenomena like belief in "fake news," conspiracist ideation, or bullshit receptivity might be more prevalent among those identifying with political conservatism, I've also pushed back on any definitive conclusions to that effect with clear counter-examples illustrating that no one political party has a monopoly on false belief. Liberals succumb to unwarranted overconfidence in their beliefs, faith in subjectivity, the Dunning-Kruger effect, confirmation bias, motivated reasoning, conspiracy theory belief, and bullshit receptivity just like conservatives do. When our bias blind spots prevent us from recognizing false beliefs within ourselves and only calling them out in our ideological opposites, we end up maligning *believers* at the expense of understanding false *beliefs*. When we fail to see the universality of false belief in this way, we fail to see its potential remedies.

With that clearly stated, our political disputes can indeed be understood according to differences in opinion about certain core beliefs and worldviews. In this chapter, I'll make good on the promise to delve more deeply into the issue of political polarization by exploring the identity-defining beliefs that lie at the heart of those differences and have given rise to a political climate that's ripe with visceral hatred. Before we get into this highly charged topic, however, we should walk back any claims—as I did with belief in conspiracy theories and bullshit receptivity—that we're living in a time of unprecedented division. Writers like Harvard Business School professor David Moss, journalist and Pulitzer Prize winner Colin Woodard, and Brookings Institution fellow Jonathan Rauch remind us that the United States has been divided throughout much of its history and that our current climate of political division pales in comparison to the Civil War or even the 1960s.[3] And the familiar proscription against discussing politics and religion in general company isn't new—it's been around at least as far back as a book of etiquette entitled *Hill's Manual of Social and Business Forms* published in 1879.

And yet, there is good evidence that liberals and conservatives in the United States have indeed become increasingly polarized over political issues during the past several decades. A 2017 Pew Poll found that the partisan divide over policy positions related to the environment, immigration, international diplomacy, racial discrimination, and government aid to the needy has widened significantly, with a marked reduction in the proportion of those holding a mix of conservative and liberal viewpoints.[4] In the 1990s, self-identified Democrats or Republicans enjoyed a considerable overlap of political opinions. Based on a measure of "ideological consistency" determined by stances on 10 different policy issues, 70% of Democrats had more liberal opinions than the median Republican, and 64% of Republicans had more conservative opinions than the median Democrat. By 2017, however, 97% of Democrats had more liberal viewpoints than the median Republican, and 95% of Republicans had more conservative viewpoints than the median Democrat. In other words, liberals have become more liberal and conservatives have become more conservative with the "ideological distance" between them growing over the past several decades, much as it had for Stephanie and Nick in just the past few years.

The growing ideological polarization of voters has seemingly followed that of politicians, with news headlines after John McCain's death in 2018 mourning not only the Arizona Senator's passing, but also the "dying art of political compromise" and the "near-extinction of bipartisanship."[5] With President Biden unable to make good on his campaign promise of using a long track record of cross-party collaboration to bridge the divide between Left and Right and his popularity ratings tanking in kind, a 2022 article in *The New Republic* declared not only that "bipartisanship is dead" but that this amounted to "great news for Joe Biden."[6] "Great news," its author argued, because it meant that Biden could abandon the futility of reaching across the aisle.

Freed from that burden, he could revert to the standard playbook of partisan politics that we've grown accustomed to ever since the late conservative commentator Rush Limbaugh—who once declared "getting along is not the objective ... defeating, politically, the people I disagree with is the order of the day"—ignited a war on compromise several decades ago.[7]

Affective Polarization

While increasing ideological polarization and an unwillingness to work toward common goals is part of the story of today's political division, research by political scientists over the past decade has revealed that there's more to it than that. A 2012 study by Stanford University political science professor Shanto Iyengar and his colleagues examined political polarization in the United States from a different angle—not from where we stand on policy issues, but from the perspective of "affect" or how we *feel* about those on the other side of the political fence.[8] Drawing from survey data spanning several decades, Iyengar's research found that the feelings of those who affiliate as Democrats or Republicans toward members of the opposing party have become increasingly negative to the point of hatred and loathing. Based on other data updated through 2019, our "thermometer rating"—how warmly or coldly we feel toward our political opponents as measured on a scale from 0 to 100 degrees—dropped from a cool 48 degrees in the 1970s, to the 30s in the 2000s, to a frigid 20 degrees as of 2019.[9]

Iyengar found that this general pattern of growing hatred aligns with other more specific metrics of "social distance" between political opposites that have also increased since the 1960s, such as the attribution of negative stereotypes—like being "close-minded," "hypocritical," "selfish," and "mean"—to those of the opposing party as well as the disapproval of one's child marrying someone from across the political aisle.[8] For example, survey data from 1960 found that only about 5% of self-described Democrats and Republicans would be "displeased" if their son or daughter married someone from outside of their own political party. By 2010, the proportion who reported that they'd be "upset or very upset" had increased to more than 30% of Democrats and nearly 50% of Republicans. More recent data from a 2020 Economist/YouGov poll found that 38% of Democrats and 46% of liberals would be upset if their child married a Republican with only 41% and 34%, respectively, responding that they wouldn't be upset at all.[10] Among Republicans and conservatives, 38% and 36%, respectively, responded that they'd be upset if their child married a Democrat with only about half responding that they wouldn't be upset at all. So, while it was a non-issue in the 1960s during a time of considerable political tension, about half of Americans who identify as Republicans or conservatives and Democrats or liberals today would at least be worried if their child married someone from the opposing political party. When explaining why, most poll respondents expressed concerns about the friction it would cause in their children's marriage based on inevitable conflict, with some characterizing interparty marriages as "impossible."

Remarkably then, we've become less tolerant of cross-party marriage as a society over the past few decades even as we've become significantly more tolerant of interracial and same-sex marriage. Meanwhile, social distance and antipathy between members of opposing parties—what's been called "out-party hate"—has become a more potent predictor of voting behavior than "in-party love."[9] A sobering reality of modern-day politics is, therefore, that we often don't vote because we support a specific candidate or a particular issue but because we don't want what our political opponents are offering. This is reflected in the oft-heard claim that voters have picked candidates in recent Presidential elections based on the "lesser of two evils"—many of us aren't so much voting for what and who we want, but rather what and who we don't want.

Recent research by University of Maryland political science professor Lilliana Mason adds to Iyengar's conclusion that affective polarization has less to do with the "what" of issues than it does with the "who" of our political identities. In a study published in 2019, Mason distinguished between two separate aspects of political ideology—"issue-based ideology," defined by what one believes about the issues, and "identity-based ideology," defined by one's party affiliation.[11] Based on her analysis of 2016 survey data, identity-based ideology—whether we identify as Democrats and liberals or Republicans and conservatives—was by far the stronger predictor of social distance, not any "set of coherent issue positions." Modern-day political polarization therefore appears to be much more about what "team" we're on than anything else, leading Mason to conclude that we've become "ideologues without issues." Other research has demonstrated that our identity-based polarization isn't limited to mere affective or emotional dislike: it even includes physical sensations of disgust and repulsion.[12] Indeed, Iyengar refers to our political leanings and preferences as more "primal" than anything rational.[13]

When it comes to political beliefs, Dartmouth College political scientists D. J. Flynn and Brendan Nyhan, along with their University for Exeter colleague Jason Reifler, have made the case that it's this kind of identity-based partisanship that's the primary driver of confirmation bias and "directionally motivated reasoning about facts" as well as the disconfirmation bias, motivated disbelief, and motivated ignorance that can lead us to mistrust our ideological opposites and embrace false or deceptive counter-narratives, as I've described throughout this book.[14] Meanwhile, University of Toronto psychologist Keith Stanovich prefers the term "myside bias" to describe what he sees as a particular subclass of confirmation bias and motivated reasoning that underlies so many political disputes today—a preference not only for information that supports what we already believe as individuals, but also what our "side" or "team" believes.[15] Indeed, a 2020 study found that higher levels of affective polarization were associated with greater belief in "in-party congruent misinformation" among both Democrats and Republicans.[16] Taken together with Mason and Iyengar's conclusion that we're ideologues without issues driven by primal instincts, this provides additional validation of what I suggested back in Chapter 3—that, when it comes to our political and other cultural beliefs, we often base them on our social affiliations rather than the other way around.

New York University professor of philosophy and law Kwame Anthony Appiah has an even simpler way of summing things up—"people don't vote for what they want; they vote for who they are."[17] "All politics," he claims, "is identity politics."

The Perils of Sectarianism

That voter's attitudes are driven more by motivated reasoning based on team affiliation than by informed opinions about policy issues helps to account for several realities of modern-day politics. For one thing, it adds to an understanding of the "social pass" we give politicians for lying and bullshitting that I discussed in the previous chapter as well as our current if long-standing tolerance of hypocrisy and ideological inconsistency in our politicians and within our political parties.[18]

Most of us who are sports fans can probably recognize this kind of inconsistency in ourselves when we react to penalty calls—if a referee fails to make a call that would have favored our team, it often seems like a gross miscarriage of injustice, but when it's our team that benefits from a non-call, we turn a blind eye. That kind of home-teamsmanship is just as recognizable in our own political "penalty calls" if we look objectively. Just like fans who spent years hating rival teams' superstars like Roger Clemons, Tom Brady, and LeBron James before embracing them when they joined the home team, we readily forgive the sins of politicians and put up with all manner of behavior that we'd otherwise find intolerable so long as they're on our side. Before Donald Trump tossed his hat into the political arena, he identified as "more of a Democrat" and "pro-choice,"[19] but evangelical conservatives who we'd expect to most take issue with such statements became some of his most fervent supporters after he won the presidency as a Republican. During the confirmation hearings for Supreme Court Justice Brett Kavanaugh in 2018, liberals put their trust in the testimony of Christine Blasey Ford and the other women who accused him of sexual assault but failed to rally behind Tara Reade when she accused Joe Biden of the same in 2020. And while Democratic leaders used to be vocal about the hazards of illegal immigration and Republican politicians once regarded Russia as a serious political threat throughout the 1980s and 1990s, those party positions have since been abandoned, with voters seemingly following along without batting an eye. Such capriciousness is so commonplace because, as Iyengar points out, the passionate and primal support we have for our political party's policy positions often occurs without any informed knowledge of the details surrounding those positions.[13]

That our political attitudes are more primal and identity-based than rational and issue-based also completes an account of why our attempts at political debate can degenerate so quickly into "flame wars" characterized by anger and incivility, as I discussed in Chapter 3. Mason's finding that identity-based ideology trumps issue-based ideology reinforces the idea that when we're driven by out-group hatred, we're unlikely to trust the claims of our ideological opponents—even when they constitute verifiable facts—much less acknowledge that common ground might exist between us. And it's not just "antisocial media" that's fanning the flames of discord. Iyengar notes that the rise in affective polarization over the past several decades has also followed in lockstep with the negativity of political campaign rhetoric and its amplification through the irresistible spectacle of political sportscasting.[13]

If we're trying to win a sports match—or, for that matter, a war—then we don't want to listen, learn, collaborate, or meet in the middle. We want what Rush Limbaugh called for—to win and see our opponents destroyed in the process. No wonder then that much of our political discourse today is framed as a battle between "us" and "them"—whether Democrats and Republicans, liberals and conservatives, Left and Right, or "blue team" and "red team"—or that the rhetoric of that discourse often amounts to an exchange of insulting labels like "libtards," "fascists," "radicals," "social justice warriors," "deplorables," "snowflakes," "idiots," "extremists," "nutjobs," "fanatics," "cultists," or "Nazis."[20] And no wonder that half of the country is declaring—wrongly as I hopefully made clear in Chapters 1 and 6—that the other half must be suffering from "mass delusion," "mass psychosis," or "mass formation psychosis."[21] Even violent political rhetoric has managed to slip into the mainstream, as epitomized in November 2021 when Republican Congressman Paul Gosar posted a video of himself as an anime character killing Democratic Congresswoman Alexandria Ocasio-Cortez. Although he was called out for the violent imagery, Gosar's office dismissed it as "about fighting for the truth" and responded that "everyone needs to relax."[22]

When pejorative name-calling that dehumanizes and "others" our political opposites takes the place of productive dialogue toward a common goal, it's easy to see how people can come to adopt the kind of conspiratorial thinking that portrays opponents—whether literally or figuratively—as "Satan-worshiping pedophiles" or "lizard people." But the potentially harmful effects of such othering extends much deeper than that. A 2020 commentary from a veritable who's who of academic experts bridging the political, psychological, and organizational sciences including Iyengar, Mason, Nyhan, and others like Northwestern University's Eli Finkel and Cynthia Wang, University of California Irvine's Peter Ditto, MIT's David Rand, and New York University's Joshua Tucker and Jay Van Bavel warned that "political sectarianism"—described as a "poisonous cocktail of othering, aversion, and moralization"—represents a grave threat to America's democracy.

> [G]iven that [political] sectarianism is not driven primarily by [political] ideas, holding opposing partisans in contempt on the basis of their identity alone precludes innovative cross-party solutions and mutually beneficial compromises. This preclusion is unfortunate, as common ground remains plentiful. Indeed, despite the clear evidence that partisans have grown increasingly disdainful of one another, the evidence that they have polarized in terms of policy preferences is equivocal.

Along the way, the causal connection between policy preferences and party loyalty has become warped, with partisans adjusting their policy preferences to align with their party identity.... Overall, the severity of political conflict has grown increasingly divorced from the magnitude of policy disagreement.

Political sectarianism consists of three core ingredients: othering—the tendency to view opposing partisans as essentially different or alien to oneself; aversion—the tendency to dislike or distrust opposing partisans; and moralization—the tendency to view opposing partisans as iniquitous. It is the confluence of these ingredients that makes sectarianism so corrosive in the political sphere. Viewing opposing partisans as different, or even as dislikable or immoral, may not be problematic in isolation. But when all three converge, political losses can feel like existential threats that must be averted—whatever the cost."[9]

The most concerning of those costs, the authors argue, is the adoption of "antidemocratic tactics when pursuing electoral or political victories" along with a "willingness to inflict collateral damage." Evidence that such costs have already been incurred among democracies around the world is mounting. For example, Yunus Orhan, a political science graduate student at University of Wisconsin-Milwaukee, analyzed data from 170 election surveys across 53 countries conducted between 1996 and 2020 to conclude that affective polarization, but not ideological polarization, has been associated with "democratic backsliding" or "the deterioration of qualities associated with democratic governance" such as free elections and equality before the law.[23] Researchers at the Carnegie Endowment for International Peace have likewise found that some kind of major democratic decline—including backsliding, democratic collapse, and transition to autocracy—has been the most common outcome of "pernicious polarization" within democracies dating back to the 1950s.[24] With no other established democracy that has remained stuck in a state of polarization for as long as we have, the United States may be increasingly vulnerable to backsliding.

Beyond the risk of democratic erosion, Louisiana State University political science professor Nathan Kalmoe has been warning us for the past decade that violent political rhetoric—even when it's metaphorical, such as when we talk about political "fights," "battles," and "wars"—can not only further fuel anger and political polarization but also pave a path to real aggression and violence.[25] Working together, Kalmoe and Mason have determined that such political violence is best predicted at the level of the individual by having an "aggressive personality" along with "strong identification with a party" and "partisan moral disengagement"—that is, seeing members of the opposing party as "evil, less than human, and a serious threat to the nation."[26]

This kind of dangerous alignment can occur not only among individuals, but also within ideological groups, as I'll discuss in the next chapter. Indeed, with increasing affective polarization and the dehumanizing portrayal of our ideological opponents as the enemy over the past several years, we've seen more than our share of civil unrest played out in larger-scale violence. In 2020, protestors-turned-rioters set fire to police stations and other government buildings in Portland and Seattle, while in 2021, thousands stormed the US Capitol with some chanting "Hang Mike Pence!" and planning to take politicians hostage. In the wake of those chaotic events, a sizeable majority of Americans have become worried about the potential for more political violence in the future.[10] But while such concern seems appropriate, it's been accompanied by something more worrisome—an increasing acceptance that violence might be justified. In 2010, polls found that only about 16% of Americans condoned political violence under certain circumstances—that proportion inched up to 23% in 2015 and rose to as high as 34–40% between 2020 and 2022.[27] With political violence increasingly regarded as both justified and necessary to protect against perceived threats, it's not hard to imagine that America could be on a path to more violent conflict or even another civil war.

The Politics of Race

Some years ago, Jane Elliott, a former American elementary school teacher turned anti-racism activist, was giving a lecture to a large and mostly White audience. She told them, "If you as a White person would be happy to receive the same treatment that our Black citizens do in this society, please stand." When no one stood, she said, "You didn't understand the directions. If you White folks want to be treated the way Blacks are in this society, stand!" When still no one budged, she concluded, "Nobody is standing here. That says very plainly that you know what's happening. You know you don't want it for you. I want to know why you are so willing to accept it or to allow it to happen for others."

Her point, the audience was slow to figure out, was that they should've stood if they believed—hypothetically or ideally—that White and Black people should be treated the same way. But since they knew that such racial equality didn't exist, the audience was giving in to the status quo by staying in their seats.

While many of the political scientists mentioned in this chapter have claimed that affective polarization has eclipsed ideological polarization here in the United States, with ample room to find middle ground if we only tried, the 2017 Pew Poll that I mentioned earlier tells something of a different story if we look more closely at issues related to race.[4] Ideological polarization between voters who identify as Democrats and Republicans has indeed been sizeable and growing, especially over agreement with these three statements:

- "Blacks who can't get ahead in this country are mostly responsible for their own condition" (53% of Democrats and 66% of Republicans agreed in 1994, 28% of Democrats and 75% of Republicans agreed in 2017).
- "Racial discrimination is the main reason why many Black people can't get ahead these days" (39% of Democrats and 26% of Republicans agreed in 1994; 64% of Democrats and 14% of Republicans agreed in 2017).

- "The country needs to continue making changes to give Blacks equal rights with whites" (57% of Democrats and 30% of Republicans agreed in 1994; 81% of Democrats and 36% of Republicans agreed in 2017).

Across these issues, the gap between Democrats and Republicans has increased dramatically from nearly two- to fourfold since the 1990s. Other data from 2020 tells a similar story of significant partisan polarization around racial issues today. A poll by Democracy Fund + UCLA found that 77% of Democrats and 45% of Republicans believed that Blacks experience a "great deal" or "a lot" of discrimination, a CBS News poll found that more than 80% of Democrats and just over 20% Republicans believed that "White people have a better chance of getting ahead in today's society than Black people," and a Yahoo!YouGov poll found that 83% of Democrats and 39% of Republicans believed that there is "a problem with systemic racism in America."[28]

What's most striking about these numbers is that, as a general rule, partisan affiliation has become a bigger predictor of ideological disagreement around these issues than any other demographic difference including age, gender, level of education, or even the racial identity of respondents. This suggests that the story of political polarization in America today can't be properly told without shining a light on the convergence of party identity and ideological identity around issues of race. According to Mason, it's just this kind of convergence that's most likely to stoke bias, incite anger, and activate violence between opposing political groups.[29]

Implicit Bias

Just what is it that's behind political polarization over racial issues today? In psychology, social identity theory points us toward an answer. Tribalism—our tendency to sort ourselves into "us" and "them" or "my team" and "their team"—represented an ancient evolutionary survival advantage before communal living within larger civilizations and altruism helped us overcome the challenge of competition for limited resources.[30] Today, such vestigial tribalism manifests as ideological and affective polarization based on in-group favoritism and out-group prejudice and discrimination that continues to plague multicultural societies and international politics. According to social identity theory, such prejudices exist in the form of more explicit, overt, and conscious prejudice—like bold-faced racism—as well as more implicit, hidden, and unconscious biases.

Over the past few decades, with a decline in overt racism following the abolishment of policies like segregation following the Civil Rights Movement of the 1960s, implicit biases about race have taken center stage in providing an account of why racial disparities persist in places like the United States. Implicit bias can be defined as a kind of cognitive bias related to the automatic attitudes and stereotypes we associate with almost anything from groups of people to nonliving things like guns, drugs, or Brussels sprouts. In 1998, psychologist Anthony Greenwald and his colleagues at the

University of Washington published the results of a study involving the newly developed Implicit Association Test (IAT) that measured implicit biases through the strength of associations between a wide range of semantic or word categories including racial surnames.[31] Since then, the IAT has been adapted to measure a variety of more specific implicit biases, most notably those related to our automatic associations with different racial groups.

The Black-White Race IAT tests for reaction times related to the pairing of photographs of Black and White faces with "good" words like delight, pleasure, triumph, enjoy, glad, cheer, happy, and terrific and "bad" words like annoy, poison, bothersome, yucky, awful, despite, horrific, and selfish. During one part of the test, the test-taker is instructed to press a button when either a White face or a good word appears and to press another button when a Black face or a bad word appears. During another, the pairings are reversed, with test-takers instructed to press a button when a White face or a bad word appears and another button when a Black face or a good word appears. When response times are slower or contain more mistakes when Black faces/good words or White faces/bad words are paired compared to when White faces/good words and Black faces/bad words are paired, this is quantified as having an "automatic preference" for European Americans (Whites) over African Americans (Blacks) or vice versa.

If that sounds confusing, there's no better way to understand the IAT than to just take the test oneself. It's easy to do since it's been readily available online through Harvard University's Project Implicit since at least 2000. In 2007, the results of more than 2.5 million tests collected over six years were published, with the main finding that automatic preferences for Whites over Blacks were detected among 68% of all respondents with only 14% demonstrating the reverse preference pattern.[32] When the results were compared across different racial groups, only Blacks demonstrated a preference for Blacks over Whites. Although those implicit automatic preferences were correlated with self-reported explicit racial attitudes in the sample, that association has not been replicated in other studies so that implicit bias for Whites over Blacks is typically modeled as something meaningfully distinct from overt racism. Nonetheless, it's been claimed to be an important source of persistent racial disparities across various sectors of society including education, healthcare, policing, and criminal justice as well as a potential cause of racial violence. Such claims suggest that while explicit or overt racism may have receded through the years, implicit or unconscious racial preferences remain both pervasive and impactful. In other words, we're all at least a little bit racist and even that little bit can result in discrimination that can be profoundly harmful.

Since there's significant disagreement between partisans over whether Blacks are victims of racism and whether systemic racism even exists in the United States, it shouldn't be surprising to hear that there's also significant disagreement and debate over the IAT. Critics have argued that the IAT's psychometric properties, such as its construct validity (its ability to measure what it claims to measure) and test-retest reliability (how well scores are consistent across multiple tests taken by the same individual), are poor such that there can be significant variation of IAT scores depending on context and that it isn't a very good tool for predicting discriminatory behavior at the level of the individual.[33] Proponents of the IAT—including its creators—have countered that there's plenty of evidence to support the test's validity, reliability, and predictive utility.[34]

Since it's well beyond the scope of this chapter to hash out those claims and counterclaims to declare a winner, I'll instead focus on one particularly counter-intuitive finding that can help us better understand just what the IAT tells us. To do that, let's take a look at the results of a 2015 Pew Research Center study that administered the IAT to 3,029 US respondents who identified as either White, Black, or Asian.[35] In that sample, 48% of Whites had an IAT result consistent with an unconscious preference for Whites over Blacks and 50% had a result consistent with an unconscious preference for Whites over Asians. No surprise there. But 25% of Whites had a preference for Blacks over Whites, 19% had a preference for Asians over Whites, and 27–30% had no preference at all. Among the minority groups, while 45% Blacks had a preference for Blacks over Whites, 29% had a preference for Whites over Blacks. Likewise, although 42% of Asians had a preference for Asians over Whites, 38% had a preference for Whites over Asians. So, while Whites, Blacks, and Asians were most likely to prefer their own race over others, as we might expect, the proportion that did was less than the majority of each racial group. Furthermore, among the racial minority groups, the difference in the proportion of those who had automatic preferences for their own race over Whites versus preferences for Whites over their own race was relatively small.

That a sizeable proportion of racial minorities—29% Blacks and 38% of Asians—seem to have counterintuitive preferences for Whites has led psychologists like B. Keith Payne from the University of North Carolina at Chapel Hill to question whether IAT responses really represent the unconscious attitudes of individuals.[36] According to what he calls the "bias of crowds" model, he instead argues that implicit biases represent a kind of global awareness of cultural biases and systemic racism that he calls a "social phenomenon that passes through individual minds, rather than residing in them."[37] In other words, maybe almost a third of Blacks have automatic "preferences" for Whites because they know all too well that Whites enjoy privileged status in the United States. After all, there's even a semantic association between the words "white" and "good" or "clean" and "black" and "bad" or "dirty" independent of race or people's faces.

While Payne's explanation has a certain exculpatory appeal, New York University professor of psychology, politics, and data science John Jost doesn't think that individual bias can or should be so neatly separated from more pervasive sociocultural stereotypes. Jost defends the validity, reliability, and utility of the IAT and contends that the finding of implicit biases among minority groups against their own race more likely represents a kind of individual "self-hatred" that has been repeatedly detected over 75 years of social science research not only among racial minorities but also among those who are poor, obese, or gay who demonstrate automatic preferences for those who are rich, non-obese, or straight.[38] Jost believes this can be accounted for based on "system justification theory," whereby existing status quos are preserved and legitimized through the explicit and implicit attitudes of both privileged groups (as we would expect) but also disadvantaged groups within societies where the social mobility of those groups is limited. For advantaged majorities, implicit biases that are decoupled from overt racism reflect an unconscious desire to preserve privilege without guilt, whereas, for disadvantaged minorities, they can act as a kind of emotionally protective denial that discrimination exists or is unjust.

If we wish to resolve these two competing perspectives, we should start by acknowledging that Payne's "bias of crowds" model and Jost's "systemic justification theory" aren't that far afield from each other. Both models agree that implicit bias reflects existing systemic bias—their difference hinges on the extent to which individual biases exist and can really be disentangled from cultural biases. By extension, the two models also diverge on just how much we should own up to our implicit biases. Few of us are eager to acknowledge that we might be harboring the kind of unconscious racial prejudices that researchers like Jost claim are revealed by the IAT. Indeed, when many of us—including some of the IAT's original developers—complete the IAT and see the result, our reaction is often one of shock, embarrassment, or denial.[39] Whereas Payne's model allows us to blame culture at large, Jost's doesn't let us off the hook so easily. After all, Jost argues, how can cultural stereotypes persist if they aren't held internally by at least some individuals?

A reasonable synthesis is that there is no "one-size-fits-all" model to characterize individual IAT results and implicit biases. Just how much we absorb and internalize cultural attitudes about race—and more importantly what we do about it—will vary from person to person in a way that may or may not be reflected in our IAT scores. And, either way, our unconscious biases don't always carry over into our conscious beliefs and explicit attitudes or predict discriminatory behavior. But that imperfect connection between bias, belief, and behavior doesn't mean we should let ourselves get away with denying that we have any individual implicit biases and that they might negatively impact the lives of others.

Now that we have a better understanding of what the IAT might tell us, we can bring the discussion back to the ideological polarization between Democrats and Republicans around race. Based on converging lines of evidence, one of the best predictors of whether or not we have implicit biases about racial minorities—and how much we acknowledge them—is partisan identity. In the study of 2.5 million IAT responses collected between 2000 and 2006 that I mentioned earlier, both implicitly measured and explicitly stated preferences for Whites over Blacks were detected at a significantly greater rate among conservatives compared to liberals.[32] Given that partisan difference, along with the correlation between implicit bias and explicitly stated preferences in that large sample, we can infer that the ideological polarization around race between Democrats or liberals and Republicans or conservatives detected in polls does indeed reflect a greater degree of individual and personal racial bias among conservatives as opposed to any greater awareness of cultural attitudes.

As further evidence, this conservative bias holds true not only among racial majorities, but among racial minorities as well.[38]

Together, these findings support the system justification model which claims that, among conservatives, implicit biases about race as well as explicit ideological beliefs about racial issues represent both conscious and unconscious desires to keep the existing social order in place.[40] In other words, just as the word "conservative" suggests, a major underlying motive of conservatism involves resisting change and preserving the status quo—and, in the case of racial majorities, the privilege—of discriminatory policies that are woven into the fabric of society. And as with conservatives' denial of climate change that I discussed in Chapter 6 and will revisit in the next chapter, motivated denial is often about resolving cognitive dissonance by refusing to believe inconvenient truths. Accordingly, it may be that those inclined to refute that the IAT reflects individual bias or that systemic racism exists have the most to hide—or lose—from acknowledging the opposite.[41] In other words, as Shakespeare suggested, such individuals doth protest too much.

Identity Threat

During the protests taking place across the United States during the summer of 2020, a friend of mine who is Latina, doesn't identify as either Democrat or Republican, and has family members in law enforcement mentioned that while she believed that George Floyd's death was emblematic of Blacks being victims of systemic racism, she objected to the slogan "Black Lives Matter." When she heard that phrase, she couldn't help but hear it as "*only* Black lives matter" so that she felt sympathetic to those who were countering with "blue lives matter" or "all lives matter." I tried to explain that in a society where systemic racism exists—as she claimed to believe—that "Black Lives Matter" meant that Black lives should and do matter, too. When she stuck to her guns, I offered the analogy that celebrating Mother's Day doesn't mean that only mothers matter. And when kids ask why there's no Children's Day, we often remind them that "every day is Children's Day." But no matter how I framed it, I couldn't convince her.

My friend wasn't alone: an Economist/YouGov Poll from June 2020 found that 17% of Hispanics and 21% of Independents had a negative association with the "Black Lives Matter" slogan. But in that poll, the strongest predictor of how respondents perceived the slogan wasn't race, but once again partisanship—77–83% of Democrat and liberals perceived it as positive with only 4–6% perceiving it as negative, whereas 44–46% of Republicans and conservatives perceived it as negative with only 25% perceiving it as positive.[42]

Why is it that my friend could acknowledge that systemic racism exists but could only hear "Black Lives Matter" as "yours doesn't?" As with implicit bias, social identity theory provides an answer by noting that tribalism—along with its modern manifestations within explicit attitudes and implicit biases about racial out-groups—is rooted in threat perception and fear. We circle the wagons and draw imaginary borders around our "own kind"—whatever that might mean to us—because we're afraid that those who we "other" are going to take what's rightfully ours, not only in terms of resources and material goods, but perhaps even our lives. Social identity theory therefore suggests that partisan differences in the perception of slogans like "Black Lives Matter" as well as in statements like "racial discrimination is the main reason why Black people can't get ahead these days" or "the country needs to continue making changes to give Blacks equal rights with Whites" are determined by concerns about existential threats that out-groups pose to in-groups. This conclusion matches other results from the 2017 Pew Poll that I cited previously: in addition to ideological polarization over racial discrimination toward Blacks, Democrats and Republicans have also become more deeply divided over whether "immigrants strengthen the country because of their hard work and talents," with 84% of Democrats and only 42% of Republicans agreeing, compared to 32% and 30% of both Democrats and Republicans, respectively, in 1994.[4]

Thinking about existential threats to our racial identities goes a long way to understanding politics in the United States over the past two decades as we've transitioned from electing the first-ever Black president in Barack Obama to electing Donald Trump on a platform of nationalism. In 2016, in the wake of the Obama's two-term presidency, Trump's win took a lot of people by surprise. With Hillary Clinton leading in the polls down the final stretch of the election, many liberals were taking another Democratic victory for granted. Even Trump himself—if we are to believe his personal lawyer Michael Cohen—wasn't expecting to come out on top.[43] And so, when Trump did emerge the victor, there was a good bit of scrambling on the part of political pundits to find clues that would explain what they'd missed. One early theory was that Clinton had lost the election for the same reason that her husband had won in 1992—the economy. According to the "left behind theory," Trump's victory was a victory of the working class—the downtrodden voters of "flyover" country who'd lost their jobs to factory closures and the global outsourcing and whom Clinton had failed to court. To them, "Make America Great Again" resonated as a rallying cry that heralded the reclamation of America's place as a leader in manufacturing and trade along with the hope that their jobs would be restored and their communities revitalized.

But, in short time, a closer look at the polling data revealed evidence to support a different theory. Trump's win hadn't been delivered by the working class and the downtrodden—in fact, Clinton had beaten Trump among voters making less than $50,000 a year. But she did lose among White voters in that income bracket, just as she lost among White voters in every income bracket. She even lost, albeit narrowly, among White women. Clinton hadn't therefore been defeated because of the economic woes of the blue-collar workers out of jobs; she'd lost because voters who'd white-knuckled it through Obama's two terms felt increasingly threatened by growing talk of the need for not just racial equality but racial equity that would level the playing field for minority groups and the specter of wealth redistribution through Democratic Socialism, as well as census data predicting that racial minorities would become a majority by 2045. So, the appeal of "Make America Great Again" wasn't just about restoring America's place in the global economy: it was about resisting "cultural displacement" by preserving the privileged majority status of Whites in America.

Prior to the surprise outcome of the 2016 election, the idea that economic anxieties had become intertwined with racial resentment in the wake of Obama's eight years in office had already appeared in Jamelle Bouie's *Slate* article "How Trump Happened" as well as Derek Thompson's "Donald Trump and the Twilight of White America" in *The Atlantic*.[44] Once Trump took office, "status threat theory" gained greater footing over "left behind theory," as reflected in similar articles in *The Atlantic*, *The New York Times*, and *Vox*.[45] And this wasn't just theorizing on the part of the so-called liberal media; it was an explanation backed by research evidence.

For example, Northwestern University professors of psychology Maureen Craig and Jennifer Richeson published a series of experiments in 2014 that examined how exposing survey respondents to information about racial minorities becoming the majority in America—a "majority-minority shift"—could influence partisanship and partisan positions.[46] In one experiment, politically independent White respondents were more likely to report leaning to the Republican Party if they were asked if they'd heard that minority groups had become the majority in California, with the strongest association found among those living in the West, closer to the alleged threat. In other experiments, White respondents who were exposed to information about the anticipated majority-minority shift on a national level were more likely to endorse conservative policy positions both related to race—like opposing affirmative action and immigration—and those unrelated to race. Craig and Richeson concluded that, for Whites, the existential threat of becoming a minority is associated with a "conservative shift" in political ideology.

Meanwhile, to better understand why Mitt Romney lost the Presidential election in 2012 and Donald Trump won it in 2016, University of Pennsylvania political scientist Diana Mutz compared national survey results between those two election years.[47] Compared to 2012, changes in both candidates' and voters' positions about trade and the threat posed by China meant that the average voter ended up situated closer to the perceived positions of the Republican candidate on those issues relative to the Democratic candidate in 2016. In addition, a rise in perceived threats to both racial and global status along with support for "group dominance" or "hierarchy over equality"—but not losing a job or income or perceiving oneself as less well-off financially—predicted swing voters switching parties to vote for Trump in 2016. Support for Trump was also predicted by the perception that "high-status groups" including Whites, men, and Christians were experiencing more discrimination than "low-status" minority groups.

In much the same way, research by Duke University professor of political science Ashley Jardina found that both the racial resentment of Whites toward out-groups as well as White in-group identity—how much one's own racial identity is rated as important—figured heavily into evaluations of presidential candidates and corresponding voting behavior in 2012, 2016, and 2018.[48] A 2017 PRRI/*The Atlantic* poll similarly found that support for President Trump among White working-class voters was associated with feeling like "strangers in their own land" as well as beliefs that growing numbers of immigrants were "threatening American culture" and that "reverse discrimination" had become as big a problem as discrimination against Blacks and other minority groups.[49] Finally, an analysis by University of Chicago political science professor Robert Pape and the Chicago Project of Security and Threats found that while the 377 individuals charged for their part in the January 2021 Capitol insurrection came from both "red" and "blue" states, they were significantly more likely to come from counties where there was a decline in the non-Hispanic White population between 2015 and 2019.[50] Survey results from the same research group likewise found that believing that the 2020 election was stolen and a willingness to engage in violent protest was most predicted by the fear of the "Great Replacement"—the notion that Blacks and Hispanics are overtaking Whites in the population and will have more rights than Whites in the future.

Collectively, such research provides a compelling link between Republican and conservative partisan identity and White racial identity, fueled by perceived existential threats posed by both Blacks in the United States and immigrants from abroad that have influenced partisan policy positions and voting behavior over the past few decades. Indeed, over that time span, conservatives have become more worried about Muslim terrorists, immigrants from south of the border, and "entitlement programs" benefiting non-Whites in the United States while liberals have become more supportive of racial equity and the value of ethnic diversity. If we want to understand the story of political polarization in the United States—including affective polarization, ideological polarization, and the "polarization of reality," whereby liberals and conservatives can no longer agree on facts—we can't ignore the relevance of racial divisions between "us" and "them" as well as those that support "them" and those that don't.[51] Put another way, when it comes down to it, political polarization in the United States amounts to a dispute over its national identity and just who exactly "us" and "them" really is or should be.

Left, Right, and Center

For nearly two and a half centuries, the American Dream has promised that "all men are created equal" and that, from that universal starting line, we all have an equal shot at achieving happiness and prosperity within a meritocracy that rewards talent and hard work. Today, political polarization has become a tale of two very different takes on that promise. In one version, Democrats and liberals advancing "critical race theory" tell us that the dream is a fraud of systemic racism that should be corrected by leveling the playing field through programs like affirmative action, social welfare, and other initiatives falling under the broad umbrella of "diversity, equity, and inclusion." In the other version, Republicans and conservatives tell us that the American Dream remains intact but that equality is being threatened by "woke" efforts—and even a deliberate plot according to "great replacement theory"—to replace White people at the head of the line with Black and Brown Americans and new immigrants from other countries. Recalling University of California, Berkeley psychologist Paul Piff's demonstration of the illusion of control that I mentioned in Chapter 2, the conservative version insists that those at the head of the line are only there because of their hard work, not because they've been afforded any advantage or privilege. This explains why so many conservatives were incensed when President Obama declared in a 2012 campaign speech that, "Somebody helped to create this unbelievable American system that we have that allowed you to thrive. Somebody invested in roads and bridges. If you've got a business, you didn't build that."[52]

If today's political polarization can be distilled down to these two competing views of reality—with critical race theory on the Left and replacement theory on the Right—how can those perspectives be reconciled so that the divided states of America can be brought closer together again? If we ever hope to find an answer to that question, we must start by reframing the problem so that we can reorient ourselves in a more productive direction. First and foremost, we should take a more critical look at claims about psychological differences between liberals and conservatives. John Jost, the New York University psychologist who argues that system justification theory helps to explain political polarization around race, reviewed nearly 300 research studies involving almost half a million research participants and concluded that there's ample evidence for the "ideological asymmetry hypothesis" that claims that liberals and conservatives can be distinguished based on quantifiable differences in the kind of psychological needs and cognitive quirks that I've discussed throughout this book.[53] Like University of Kent psychologist Karen Douglas does when she groups the psychological needs associated with conspiracy theory belief, Jost categorizes the needs that determine partisanship as "epistemic, existential, and relational" including needs for certainty, closure, and control—the three C's—that I mentioned in Chapter 6, perceptions of and reactions to threats as I've discussed in this chapter, and the balance of intuitive thinking and cognitive rigidity versus cognitive reflectiveness and analytical thinking that's been a recurring theme throughout this book, as well as dogmatism (defined as the sense that "to compromise with one's political opponents is dangerous because it usually leads to the betrayal of our own side"). Because conservatives tend to have greater needs for certainty and control, greater concerns about threats, more cognitive rigidity and a greater tendency to think intuitively than analytically, and more dogmatism compared to liberals, this account of ideological asymmetry is sometimes referred to—with the tone of a pejorative—as the "rigidity of the right" hypothesis.[54]

There's little reason to deny that such group variances in psychological traits might explain why, as Jost puts it, "people on the right favor tradition and hierarchy, whereas people on the Left favor progress and equality."[55] But, in acknowledging those differences, we should drop the language of affective polarization that amounts to name-calling or characterizations of psychopathology epitomized in a 2016 online article that went viral with its provocative title, "A Neuroscientist Explains What May Be Wrong with Trump Supporter's Brains."[56] We should also bear in mind the refrain that I keep coming back to in this book—that we're talking about average quantitative differences rather than qualitative absolutes that are irreconcilably distinct. It's not as if only conservatives ever fall victim to the cognitive foibles of ideology and false belief—far from it. In addition, the claim that conservatives are more vulnerable than liberals to motivated reasoning and partisan bias—defined as "a general tendency for people to think or act in ways that unwittingly favor their own political group or cast their own ideologically based beliefs in a favorable light"—has been disputed by University of California psychologist Peter Ditto based on a meta-analysis of 51 studies involving more than 18,000 participants finding no such asymmetry.[57] Other research has likewise found no evidence of any Left–Right difference in intolerance toward ideological opposites or groups perceived to be threatening.[58] Finally, some studies have found that cognitive rigidity is more associated with "extremism" on both sides of the political fence rather than conservatism per se, providing evidence for an alternative "ideological extremity" or "rigidity of the extreme" hypothesis.[54] Acknowledging such findings, Jost concedes that while he finds plenty of evidence to conclude that ideological asymmetry is a reality, the imbalance is in no way "immutable, unbridgeable, or unaffected by historical and cultural dynamics."[55]

In framing the issue of political polarization, we should therefore take care not to succumb to "all-or-none" or "black-and-white" thinking that claims that all liberals are "this" or that all conservatives are always "that," whether we're talking about their psychological needs, cognitive characteristics, or ideological positions. Doing so represents a pitfall of a cognitive bias called "binary bias" that refers to our short-sighted tendency to put things and people into simplified, categorical, and dichotomous boxes.[59] In other words, it's just another kind of unproductive tribalism. The reality is that not all conservatives are resistant to change or fretting about the replacement theory, not all liberals are progressive activists and social justice warriors, and, as we've discussed, implicit biases and dominant group preferences are shared among both racial majorities and minorities alike. A more enlightened and helpful framing therefore steers clear of splitting the world into "black" and "white" in ways that fan the flames of tribalism in favor of acknowledging the many colors or shades of gray that lie in between. Too often, discussions of political polarization—admittedly including much of this chapter thus far—fail to acknowledge the substantial ideological and political diversity that exists in the United States and in the world. Not everyone is a Democrat or a liberal or a Republican or a conservative—there's a much broader spectrum of centrists, independents, moderates, populists, and political apathists in the middle along with communists, socialists, libertarians, fascists, and anarchists at the extremes—with Americans increasingly carving out identities outside of the traditional two-party system.[60]

Similarly, when we're talking about the politics of race, we should remind ourselves that foreign actors like Russia have vested interests in stoking the flames of racial discord in the United States, as I mentioned in Chapter 5.[61] That doesn't mean that we should ignore racial inequity or pretend it doesn't exist, but we should be more mindful of nuance when we separate people into "Black" and "White." For racial minorities, identifying with their minority status can be both a celebration of their cultural heritage as well as an acknowledgment of how others see and treat them. For racial majorities, counter-claims of "color-blindness" often represent motivated denial that attempts to whitewash the reality of racial discrimination.[41] That said, multiracial identities are becoming increasingly common. Obama wasn't only our first Black president, just as Kamala Harris wasn't only our first Black vice president: Obama's mother was White and Harris's was Indian. While Whites might not be the majority come 2045, neither will Blacks, Hispanics, Asians, or any other single racial group. Instead, we can expect a more multiracial and multicultural nation with more and more people checking off multiple boxes on census forms—or none at all—to describe their racial identity. Instead of fretting about majority-minority shifts, we should remember that the US motto is "E Pluribus Unum" or "out of many, one." Despite claims to the contrary, America's success as a democracy has always been as an experiment in diversity, not homogeneity.

A More Perfect Union

With these "rules of engagement" established, we can now return to thinking about Stephanie and Nick as a metaphor for America's polarization from the perspective of a marriage counselor. Much like a democracy, a healthy marriage involves two people joining together as one while still maintaining their separateness and individuality. The goal of marital counseling or couples therapy isn't necessarily to preserve that union; it's to help a couple figure out where things have gone awry and what they want to do about it. Sometimes that will mean that one or both parties will decide to divorce. If both sides do want to save their marriage, it requires that each makes a conscious commitment to do the work necessary to preserve what it is that brought them together in the first place and what they still value in it.

Whether we're talking about a marriage or a democracy, it isn't disagreement or conflict that threatens a union. A healthy partnership should be able to tolerate disputes stemming from different points of view while still working together toward common goals. In politics, there should be plenty of room for reasoned debate in the name of finding middle ground on issues related to economic policy, immigration, gun control, racial equity, abortion, healthcare, social welfare programs, or foreign relations. But when we come to view our partners as enemies unworthy of understanding, negotiation, or compromise and instead end up fighting for fighting's sake, alliances fracture and differences become irreconcilable. When a husband and wife find themselves only pushing each other's buttons, moving farther away from each other, refusing to take each other's needs into account, and unwilling to compromise, they're only left with two choices—to remain miserable or to watch their marriage end. Just so, whether we're talking about voters, politicians, or political parties, when one's strategy is only focused on shifting the Overton Window—the boundaries of what's acceptable to the public—by pushing the middle closer to an extreme, any chance of both sides coming out satisfied goes out the window. The same holds true for obstructionism, exploiting back-and-forth party majorities to railroad legislation through Congress or via executive order, and stacking the Supreme Court with partisan Justices. The endgame of such refusal to compromise or collaborate per the playbook of Rush Limbaugh is that the Left moves farther left, the Right moves farther right, or both, until the only way to win is through a one-sided victory that replaces democracy with authoritarianism.

Pledging not to move farther apart doesn't always mean that the solution—or the truth—lies in the middle of a dispute however. In Stephanie and Nick's marriage, it was Nick who'd shifted the Overton Window, moving farther to the right and venturing into conspiracy theory territory. I wouldn't therefore expect or suggest that Stephanie move farther to the right with him anymore than I'd recommend that she compromise if Nick was demanding that they have an open marriage against her wishes or was subjecting her to physical abuse. While marriage counselors generally try to avoid taking sides whenever possible, sometimes one spouse is right and the other is wrong. In such cases, it's necessary to heed the warnings of the late Nobel Peace Prize winners Desmond Tutu and Elie Wiesel who both warned that, in situations of injustice, neutrality only helps the oppressor, not the victim. Helping two sides work collaboratively toward common goals shouldn't therefore be conflated with endorsing false equivalencies and condoning unwarranted "bothsidesism."

It could be argued that a marriage isn't the best metaphor for a democracy. After all, the two sides of today's polarized America didn't exactly choose to be together. But according to the Hidden Tribes of America project—organized by More in Common, an "international initiative to build societies and communities that are stronger, more united, and more resilient to the increasing threats of polarization and social division"—an "exhausted majority" comprising as many as two-thirds of the US population is fed up with political polarization fueled by the voices of relative extremists and does support finding common ground.[62] Based on a 2020 poll from More in Common/YouGov, 82% of the 2,000 respondents believed that Americans have "more in common than what divides us" and 90% believed that "we're all in it together"—an increase from 63% in 2018.[63] That's a small sample size, but it offers some reason to be hopeful that polarization could yet recede, with the United States living up to its name by becoming more united than divided.

Still, wanting something is a world apart from achieving it. When marital counseling does help a couple find the common ground that allows them to come back together again, it often involves less arbitration of who's right or wrong and more listening to each side to understand why the perceptions of their marriage and each other are so different. In politics, that means getting to the heart of the motivated reasoning—as I've attempted to do in this chapter—that lies beneath affective and ideological polarization. But repairing a relationship that has come to be dominated by fighting also requires replacing conflict with other collaborative experiences that allow space for a couple to enjoy each other's company again. In keeping with this premise, recent research by University of Koblenz-Landau psychologist Emily

Kubin and her colleagues at the University of North Carolina at Chapel Hill and the Wharton School of Business found evidence to support that political divides are better bridged—through mutual respect if not agreement per se—by sharing personal experiences or "subjective anecdotes about lived experiences" than disputing facts.[64] "Experiences," Kubin says, "seem truer than facts."

And so, I'll conclude this chapter with the personal experiences of two people who are worth bringing into our collective consciousness. The first comes from 2016, when Ken Stern, a self-described liberal and former CEO of NPR, set out across the country to better understand "conservative America." Along the way, he "went to evangelical churches, shot a hog in Texas, stood in pit row at a NASCAR race, and hung out at Tea Party meetings," and he came away with the conclusion that Americans aren't as divided as we might think. Detailing his journey in a book entitled *Republican Like Me: How I Left the Liberal Bubble and Learned to Love the Right*, Stern wrote that although he encountered some "less than attractive types along the way," he was "almost always able to find more points of agreement and commonality than [he] thought was possible."[65] At the end of his cultural excursion, Stern denounced his Democratic affiliation and declared himself an Independent. If we hope to bridge America's political divide and find middle ground, the value of such face-to-face interactions—especially in a world where we spend so much of our social lives online—is vital to understanding our ideological opposites, reducing affective polarization, and walking us back from the brink of violent conflict.

The other personal experience worth bearing in mind comes from 1959, when John Howard Griffin—a White man like Stern—set out on a similar journey. Unlike Stern, however, Griffin shaved his head and worked with a dermatologist to darken his skin to the point of being able to pass as Black before embarking on a six-week road trip across the segregated Southern United States. As he did, he documented the overt racism that he endured firsthand in the form of physical threats and treatment as a "tenth-class citizen" along with the more subtle "silent language" of "hate stares" in a memoir entitled *Black Like Me*.[66] Griffin's experience was radically different from Stern's because he asked himself, "How else except by becoming a Negro could a White man hope to learn the truth?" In doing so, he came as close as possible to experiencing what it was like to be someone else. Of course, his experience also diverged from Stern's because it occurred nearly 60 years earlier, but before anyone dismisses his experience as antediluvian, they might want to consider repeating Griffin's experiment today. The pathway to true empathy is best tread by walking a mile in someone else's shoes. Or better yet, their skin.

9
We Are Not Our Beliefs

> Martyrdom was only ever a proof of the intensity, never of the correctness of a belief.
>
> —Arthur Schnitzler

Seekers of Truth

One crisp, fall New England afternoon during my first year in college, I was working with my roommates Mike and Adam on a "problem set"—MIT's term for homework—for our freshman physics class. The way I like to remember it, I was leading us step by step through the math on a particularly tricky question. And since Adam went on to win the Nobel Prize in Physics some twenty years later, I sometimes like to tell myself that I once helped a future Nobel laureate with his homework.

But Mike remembers it differently. He says that after I fumbled my attempt to figure out the answer, it was Adam who grabbed some chalk, went up to the little chalkboard we had in our room, and proceeded to write out the correct series of equations that took us to the solution. Given Adam's Nobel Prize—and thinking back to the fallibility of memory that I described in Chapter 2—I have to admit that Mike's version is probably a better representation of reality.

Adam Riess—now the Bloomberg Distinguished Professor and the Thomas J. Barber Professor of Space Studies at the Krieger School of Arts and Sciences at Johns Hopkins University—won the top prize in science for his postdoctoral work with the Hubble Space Telescope. While looking at light emitted from Type Ia supernovae in the distant universe, he and his colleagues observed that, contrary to the prevailing expectation that the universe's expansion that began with the Big Bang was slowing down, it was actually speeding up. Riess says that when they first saw the data, they assumed this had to be a mistake since the data didn't conform with the existing model at the time. But with repeated analyses and observations that were confirmed by another team of researchers doing similar contemporaneous work at the Lawrence Berkeley National Laboratory, they eventually realized that the data were right so that the model itself had to be revised. It's now an accepted fact of science that the universe's expansion is accelerating, most likely due to a force called "dark energy," the nature of which remains something of a mystery that Riess and other astrophysicists are still trying to unravel.

A few years ago, Riess told me that ever since he won the Nobel Prize, he's been getting a steady stream of email—as many as one or two messages a week—from random people purporting to have stumbled on some major discovery in astrophysics. About half of them tell him that he's got it all wrong, while the other half want him to validate their findings and, in some cases, even ask him to recommend them for a Nobel Prize. Either way, they seem to have what Riess calls "a deep desire to be able to say they've pulled the proverbial sword from the stone of contemporary physics." But much like my patient with schizophrenia that I described in Chapter 7 who wrongly believed that he'd worked out a mathematical recipe for cold fusion, their proposals invariably amount to a kind of scientific bullshit that Riess describes as "a word salad of terms and equations that don't follow any logic." He says it's usually obvious that those who contact him have very little formal training in physics and don't even know enough to realize they're not making sense—it's as clear an illustration of the Dunning-Kruger effect as I've ever heard. And, for what it's worth, as far as Riess can tell, everyone who has contacted him has been a man.

As we approach the conclusion of this book, this chapter begins to wrap things up by addressing how we might best go about determining what's true or false when it comes to factual beliefs while using the example of climate change denial to summarize how the intuitive mechanics of belief can so easily lead us astray. It then goes on to discuss how our most passionate opinions often consist of nonfactual ideological judgments in the form of values and morals—like the political beliefs discussed in the previous chapter—about the way we think the world should be and how other people should act. Based on the finding that it's these types of belief that are most integral to our sense of personal identity (i.e., how we define ourselves both as individuals and group members), it then concludes with a quantitative model to characterize escalating ideological commitment and the potential danger of defining ourselves according to our beliefs.

* * *

For those with a heightened need for the three C's—control, certainty, and closure—who demand definitive answers and absolute truths over ambiguity, the persistence of unsolved mysteries like "dark energy" and the constant updating of scientific theories and facts can be a source of frustration. Is the universe's expansion slowing down or speeding up? Is Pluto a planet or not? Should we eat a low-fat diet or take an aspirin once a day or shouldn't we? And if science is always changing its answers, why should we trust anything it tells us?

To address that last question, let's contrast the scientific method's process of "research" that I mentioned in Chapter 1 and that won Riess his Nobel Prize to the more instinctive way that we often react when our beliefs and expectations bump up against reality. In the late 1950s, University of Minnesota psychologist Leon Festinger coined the term "cognitive dissonance" to describe the uncomfortable psychological tension that arises when one of our beliefs conflicts with another belief or observation about ourselves, our behavior, or the world.[1] Although cognitive dissonance is now recognized as something of a household word, it's less well-known that Festinger developed the concept based in part on his study of a "UFO cult"—similar to Heaven's Gate that I described in Chapter 4—whose members referred to themselves as "the Seekers."[2] Claiming to be in communication with an otherworldly alien race that they called "the Guardians," the Seekers prophesized that humanity would be destroyed by a flood and that the Guardians would come to rescue the faithful by spiriting them away on a spaceship at 4 o'clock in the afternoon on December 17, 1954. When that moment came and went without the Guardians showing up, the Seekers didn't abandon their faith, they just moved the predicted time of the rendezvous to later in the day. And when the Guardians still failed to appear, they kept "moving the goalposts" by revising the moment of departure three more times until finally giving up on Christmas Eve. At that point, one of the Seekers—a former practicing physician—claimed that the aliens had indeed arrived, but had elected not to reveal themselves to the 200-some onlookers who'd gathered to gawk. Later, the Seekers declared that their prophesies hadn't been wrong per se, but that their faith must have spared them and the rest of the world from the apocalypse.

According to Festinger's theory and the sizeable body of subsequent research that supports it, the discomfort of cognitive dissonance means that we'll often try to do whatever we can to make it go away. But exactly how we go about doing that depends on a variety of factors including how much time, energy, and other resources we've invested in a particular belief. The more committed we've been, the less likely we are to abandon a belief. And so, the Seekers who'd given away their worldly possessions in anticipation of leaving the Earth maintained their faith in the group's dogma and kept doubling down on their predictions, whereas those less invested were able to walk away from their beliefs along with their membership in the group. As I suggested back in Chapter 4, we often rely on motivated reasoning and motivated denial to resolve conflicts between what we believe and evidence that contradicts it—cognitive dissonance means that our tendency to dismiss such evidence is proportional to the cost of giving up a cherished belief. Cognitive dissonance therefore adds another layer of explanation to understand why the 39 members of Heaven's Gate ended their lives even after the images of a spaceship hidden in the tail of the Hale-Bopp comet were revealed to be a hoax. It also provides a more complete account of why some of the QAnon faithful that I mentioned in Chapter 6 still "trust the plan" despite Trump losing the 2020 election and none of Q's other predictions ever panning out. And it underscores why those who most benefit from the injustice of systemic racism are the most likely to deny that it exists, as I discussed in Chapter 8. When motivated denial is used to resolve cognitive dissonance, the motive involves not wanting to pay the price of being wrong.

With this in mind, we can see just how much thinking like a scientist can help guide us to the truth whereas our more intuitive cognitive machinery can often lead us down a path of self-deception and false belief. Whereas the scientific method teaches us to maintain skeptical disbelief in the absence of evidence and to modify existing beliefs to fit new data, cognitive dissonance and motivated reasoning often work in the service of preserving our faith through denialism, discounting evidence so that we don't have to change our minds. This difference lies at the heart of the Dunning-Kruger effect as it applies to science. While scientific experts like Riess are constantly asking themselves "Am I wrong?," our more instinctive default is to engage in a kind of overconfident self-deception that allows us to convince ourselves—and claim to others—that we're right.

While it might seem counter-intuitive, the reason we should trust science is because of its willingness to be proved wrong. In that sense, the so-called replicability crisis in science and in psychology whereby initial experimental findings that were accepted as gospel haven't always been reproduced with repeated experiments isn't so much a crisis as it is science doing what it's supposed to do. Riess allowed new evidence, verified by repeated observations, to change his long-standing beliefs. That didn't incur a cost—on the contrary, it won him the Nobel Prize. It should be the same for all of us—modifying our beliefs based on the balance of evidence shouldn't be considered a weakness or a shortcoming: we should celebrate it as a self-correcting superpower to which we should all aspire that rewards us by steering us clear of false belief and closer to the truth.

Skepticism, Denialism, and Climate Change

As I suggested earlier, trust in science is often hampered by the common misunderstanding that the kind of skepticism and malleability of belief that's part of the scientific method means that facts and ultimate truths lie beyond our grasp. After all, science tells us that there are few absolutes and that—as I've asserted from the start of this book—beliefs are best expressed as probability judgments based on the best available evidence. And yet, we shouldn't conflate scientific skepticism with postmodernist or post-truth denialism—there's a crucial difference between the two. According to science, the truth is out there, but our understanding of it may change over time. Facts exist, though they're not necessarily set in stone. Neither of these statements mean that truth is merely arbitrary or an illusion however. Scientific facts and truths are conclusions that we draw from objective evidence based on repeated observations and controlled experiments; not subjectivity, political interests, or appeals to authority and claims of expertise alone. As physicist Richard Feynman once put it, "it doesn't make any difference how beautiful your guess is, it doesn't matter how smart you are, who made the guess, or what his name is … if it disagrees with experiment it's wrong."[3]

To that point, back in Chapter 1, I cited Atul Gawande's important distinction between trusting science as a method of uncovering the truth and trusting the word of scientists. Indeed, while science is always ready to be wrong, that's not always the case with scientists. As I hopefully made clear in Chapter 2, scientists and other experts aren't immune to overconfidence, motivated reasoning, and the other psychological pitfalls of false belief—as Feynman also quipped, "science is the belief in the ignorance of experts."[4] Even Nobel Prize winners—like Linus Pauling, William Shockley,

James Watson, Kary Mullis, and Ivar Giaever, to name a few—have been known to succumb to what's been called "Nobel disease" by going on in their post-prize careers to embrace fringe theories and spread harmful misinformation, especially outside their areas of expertise.[5] And so, trusting science should never be equated with mere "argument from authority" or blindly taking what a scientist tells us as uncontestable. Just so, Feynman advised against making claims about what "science" or a previous generation of scientists has "shown" in favor of explaining what an experiment—that is, the re-search evidence—continues to reveal to us over time.[4]

That said, scientific consensus based on such evidence shouldn't be as easily discounted as it often is. It's one thing when a lone scientist or a handful of scientists makes a claim that's debatable; it's another when nearly the entire field is in agreement. And while someone like Adam Riess might occasionally turn an entire field on its head with new discoveries worthy of a Nobel Prize, the Dunning-Kruger effect reminds us that we're not all Adam Riess, even if we might think we are. Just so, University of Minnesota professor Emily Vraga and Georgetown University professor Leticia Bode recommend more broadly that when we attempt to distinguish between reliable information and misinformation, we should do based on a combination of objective evidence and—since much of that evidence lies beyond our comprehension if we're being honest—expert consensus.[6] While that conclusion might seem eminently sensible to the point of being obvious, it's all too often ignored within a post-truth world in which expertise is dismissed as just another opinion.

Naïve Realism and the Law of Small Numbers

My good friend Scott—a now retired physician with whom I worked side by side for a decade—lives in an idyllic setting on the beach in Malibu, where he basks in the glow of sunny skies, a cool breeze blowing off the Pacific Ocean, and the sound of rolling waves coming through his window year-round. Though he likes to brag about nearly flunking out of college, he's a kind of polymath who used to enjoy regaling medical students about how the dinosaurs became extinct or how the ancient Egyptians erected the pyramids. And yet, despite a solid background in science and a well-above-average handle on meteorology, Scott doesn't believe in global warming. He's seen the data on rising temperatures and carbon dioxide emissions, but he just doesn't buy that our current warming trend has anything to do with human activity or that it represents anything beyond the same cyclic fluctuations we've been having for eons. After all, he remembers that not too long ago—in the 1970s, back when he was in medical school and I was just a kid—scientists were warning everyone about a new Ice Age.

To review why evidence and expert consensus can so easily be ignored by people in today's post-truth world, let's revisit climate change denialism that I cited in Chapter 6 as evidence of a Dark Age of Conspiracy Theories and take a closer look at how the various other psychological components of false belief described throughout this book might explain it. Starting with the facts as we know them, objective evidence tells us very clearly that average global temperatures have been rising over the past century or so, with the nine hottest years ever recorded occurring between 2015 and 2023. As I write these words at the end of 2023, we've just finished experiencing the hottest year ever recorded on planet Earth.

Since global temperatures have only been measured and recorded since the late 1800s, a longer view of temperatures going back in time requires that climate scientists reconstruct geological evidence from the trapped gas content in ice cores obtained by drilling into glaciers or the presence of isotopes detected in fossilized plankton. Studies using this kind of "proxy data" indicate that global temperatures have been cycling up-and-down for hundreds of thousands of years and that we've been riding an overall upswing of global warming over the past several millennia. These conclusions are, for the most part, undisputed.

But, based on such studies of paleoclimatology, climate scientists also tell us that the uptick in global warming has been much more drastic over just the past century, coincident with a level of atmospheric "greenhouse gas" emissions due to the burning of fossil fuels in the postindustrial era that's the highest it's ever been in human history. The conclusion that global warming is *anthropogenic* or caused by human activity in this way—as opposed to merely being part of the normal cycling of temperature due to heat from sources like the sun or volcanic activity—is based on evidence taken from plotting global temperatures against measures of solar irradiance and carbon dioxide (CO_2) levels or comparing temperatures at different layers of the atmosphere. While that might sound relatively straightforward, it involves data collection, mathematical modeling, and scientific analysis requiring a level of knowledge, specialty training, and interpretive ability that far exceeds what most of us possess.

As Vraga and Bode suggest, holding informed beliefs about climate change therefore requires that we rely on expert consensus where it's claimed that 97%—and even as much as 100%—of climate scientists agree that anthropogenic climate change (ACC) is a reality. This claim is based on reviews of the thousands of published scientific papers that have articulated a conclusion about whether climate change is caused by human activity, where 97–100% agree that it is and only 0–3% disagree.[7] Beyond those published papers, several surveys have reported anywhere from significant to overwhelming consensus on ACC among climate scientists.[8] And finally, consensus statements have been issued by numerous scientific organizations including the Environmental Protection Agency (EPA), the National Aeronautics and Space Administration (NASA), the National Oceanic and Atmospheric Administration (NOAA), and the National Science Foundation (NSF) here in the United States as well as by the United Nations' Intergovernmental Panel on Climate Change (IPCC).

Beyond the reality of ACC in the present, climate scientists further predict that continued global warming of just a few degrees will have calamitous consequences in the future if we don't curb the use of fossil fuels now. The latest IPCC report from 2022 warns us that without significant efforts to decrease CO_2 emissions across the world, we can expect not only more of the heat waves, fires, floods, rising sea levels, and ecosystem destruction (i.e., climate change due to global warming) that we've been witnessing around the world in recent years, but also an exponentially increasing risk of mass human casualty from drought, depletion of food and freshwater resources, infectious disease, and other deadly living conditions with a closing window of opportunity to avert disaster.[9]

***

Climate change denialism isn't a monolith—it can take many forms based on disbelief in any of the facts, expert assessments, and dire predictions reviewed in the preceding paragraphs. As we might expect then, determining just how many climate change deniers there are—just like trying to figure out how many conspiracy theory believers there are—depends on how we ask about the subject. For example, Pew polls from 2006 to 2012 found that 57–77% of US adults believed that there was "solid evidence the earth is warming" but only 36–47% believed it was warming "mostly because of human activity."[10] A 2021 poll by The Economist/YouGov asking about "climate change" rather than global "warming" found that 53% believed that the world's climate is changing as "a result of human activity," 24% believed that it's changing but not because of human activity, and 9% believed that it isn't changing at all.[11] Only when a 2020 Pew poll gave respondents three options to agree or disagree—as opposed to forcing them into a black-and-white binary—did as many as 81% endorse belief that human activity contributes to climate change, with 49% endorsing "a great deal" of belief, 32% endorsing "some" belief, and only 19% believing that human activity contributes "not too much" or "not at all."[12] Such findings are consistent with other research demonstrating that belief in ACC increases when poll respondents are given the chance to quantify their belief conviction beyond a "yes" or "no" answer.[13] When researchers at the Yale Program on Climate Change Communication used a yearly survey administered from 2008 to 2021 to stratify responses into six different levels of concern about climate change, the 2021 results found that 75% were either "alarmed," "concerned," or "cautious," while only 24% were "disengaged, doubtful, or dismissive."[14] With those who described themselves as "alarmed" doubling from 18% in 2017 to 33% in 2021, climate change denial appears to be on the decline, but some form of it—especially around the mostly hotly contested point of whether climate change is man-made—nonetheless persists to some degree.

Political scientists Patrick Egan from New York University and Megan Mullin from Temple University argue that one reason climate change denial won't go away has to do with the fact that our personal experience of the weather is far more convincing than reading or being told about global temperature averages. In one study, they found that for every 3.1°F increase in local temperatures above normal, Americans become one percentage point more likely to agree that there's "solid evidence that the Earth is warming."[15] Given that the hottest temperatures on record have occurred in the past nine years, it's no surprise that, as of 2021, only 10% of Americans currently deny that "climate change is happening."[16] But in another study, Egan and Mullin found that although average temperatures in the United States increased between 1974 to 2013, as many as 80% of Americans still experienced more favorable overall conditions with much milder winters and less humid summers over that time span.[17] In 2021—the sixth hottest year on record at the time—the United States also had its coldest winter in 30 years while temperatures at the South Pole dropped to the lowest ever recorded. This suggests—recalling how Senator Inhofe used a snowball to refute global warming back in 2015, as I mentioned in Chapter 6—that we're creatures of the moment, with heat waves making us more concerned about global warming and cold winters allowing us dismiss it as a statistical anomaly.

In a related fashion, since most of us experience not only seasonal changes that cycle between sweltering heat in the summer and frigid cold in the winter but diurnal temperature shifts of at least 20°F between day and night, it can be hard to grasp that a mere 2.7–3.5°F rise in mean global temperature could possibly result in the kind of catastrophic disaster predicted by the IPCC. Collectively, such impressions based on our subjective perceptions are exactly what we would expect due to the "seeing is believing" dictum of naïve realism that I discussed in Chapter 1, as well as the misleading reliance on the "law of small numbers" that I discussed in Chapter 2. Perhaps it's no wonder then that my friend Scott, who lives on the beach and rarely ventures beyond a 10-mile radius, is so nonplussed about global warming. Meanwhile, just on the other side of that border zone where I used to live in the San Fernando Valley, it regularly got to above 110°F in the summer with wildfires so common that they've become expected seasonal events.

Climate Science Versus Big Oil

Besides naïve realism and the Dunning-Kruger effect allowing some of us to turn a deaf ear to the warnings of climate change experts, there's ample evidence that climate change denial can also be attributed to the flea market of opinion and the disinformation industrial complex that I covered in Chapters 4 and 5. As of 2017, nearly 90% of Americans were still unaware of climate scientists' consensus on ACC.[18] This isn't particularly surprising since few of us have been keeping up with—or even have access to—climate science research journals, whereas an enormous amount of misinformation about climate change has been readily available over the past several decades in the form of books, television media coverage, and online sources of information. For example, a 2019 analysis by the online activist network Avaaz reported that more than 5,000 YouTube videos refuting the claims of climate scientists with titles like "The Truth About Global Warming," "What They Haven't Told You About Climate Change," and "Climate Change: What Do Scientists Say?" had been viewed by more than 21 million users due to promotion by YouTube and support from advertisers including some of "the world's most widely recognized and trusted household brands."[19] Many of these videos were created by climate change denial groups that have been working for years to give the false impression that ACC remains a topic of legitimate scientific debate with two opposing sides rather than a consensus of 97% or more. In 2019, a group calling itself Friends of Science posted a YouTube video entitled "No Climate Change Emergency Say 500 to UN" that publicized a letter denying ACC signed by "500 knowledgeable and experienced scientists and professionals in climate and related fields."[20] Despite claims and appearances, however, the greater a climate scientist's expertise as defined by active climate science research with publications in peer-reviewed journals, the less likely they are to refute ACC.[7] In other words, the self-proclaimed scientific authorities who encourage climate change denial are, as a general rule, anything but experts. And academic credentials aside, their dubious claims and interpretations of existing research are typically based on cherry-picked studies and flawed methodological analyses under the guise of objective evidence.[21] If we were to agree to count them as scientists, then we would have to qualify them as endorsing the very definition of "fringe" or "junk" science. Finally, as I suggested in Chapter 6, if we "follow the money" that's funding climate change denialism, the connections to "Big Oil" and the fossil fuel industry become all but impossible to ignore.[22]

If the battle over the credibility of ACC as a scientific fact is being fought between climate scientists and Big Oil, how do the rest of us choose sides and decide what to believe? As I've suggested throughout this book, the answer lies in examining our trust and mistrust of informational sources along with the underlying motives of our motivated reasoning. Although scientists were among the most trusted sources of informational authority in the United States, with 76–87% of Americans endorsing either a "fair amount" or a "great deal" of trust in them between 2016 and 2021,[23] a 2019 Pew poll found that 35% also believe that "the scientific method can be used to produce any conclusion [a] researcher wants."[24] Meanwhile, Michigan State University sociologist Aaron McCright found evidence in a 2016 study that climate change denialism was best predicted by trust in groups representing the "industrial capitalist system" including the "US business community" and "oil and gas companies" as opposed to trust in "environmental groups" and "the academic/scientific community."[25] In addition, with trust in the industrial capitalist system over the scientific community strongly associated with Republican party affiliation, McCright concluded that "the most robust predictor of climate change skepticism is political orientation." That claim matches 2021 poll data revealing that 77% of Democrats and 86% of liberals versus 53% of Republicans and 22% of conservatives endorsed belief in ACC, and 83% of Democrats and 94% of liberals versus 46% of Republicans and 44% of conservatives believed that "climate change and the environment" is either a "very" or "somewhat important" issue.[11]

Echoing the previous chapter's conclusions about the intersection of political identity and racial identity, McCright has also found that conservatives'—and especially White male conservatives'—support of the industrial capitalist system can be attributed to the kind of "system justifying" tendencies described by John Jost that maintain the status quo.[26] But there's also evidence to support that partisans are simply toeing their party's line so that accounting for the motivated reasoning and myside bias of individuals requires that we also illuminate the motives and vested interests of party leaders and other apex predators of the disinformation industrial complex. While climate change denialists often portray themselves as courageous contrarians pushing back against the supposed vested interests of environmental scientists and the Green Lobby, a 2013 study found that within a sample of 108 published books refuting ACC, 72% had verifiable links to conservative think tanks.[27] A more recent analysis of 128 "climate change counter-movement" groups likewise found that from 2003 to 2018, they were financially supported to the tune of nearly $10 billion, much of it in the form of "dark money" flowing from conservative political foundations like the American Enterprise Institute, the Heritage Foundation, and the Hoover Institution.[28]

Why is it that conservative think tanks and political foundations are pumping so much money into promoting climate change denial? Some international perspective sheds light on an answer. Across a 2019 sample of 23 "rich countries" around the world, the United States had the third highest proportion of climate change deniers, second only to Saudi Arabia and Indonesia, and the largest proportion who believe that climate change is a hoax.[29] In addition, the Left–Right partisan divide over climate change concern is much bigger in the United States than it is in many other European countries, South Korea, or New Zealand.[30] Viewed through this global lens, it's hard not to see the relative prominence of climate change denial in the United States as linked to the fact that it's the largest oil producer in the world. If we also consider the financial ties between conservative political leaders in the United States and Big Oil that I mentioned in Chapter 6, we can better appreciate how conservatives aren't really "anti-science" across the board as is sometimes claimed. Rather, as McCright's research has shown, conservatives instead tend to support and trust the "production scientists" who have worked toward innovations supporting economic growth over the past several decades while opposing the "impact scientists" who have communicated inconvenient truths about the potential harms of economic production.[31]

In a similar way, other researchers have attributed ideologically motivated climate change denial to "solution aversion," whereby the costs of trying to stop global warming—such as those related to curbing fossil fuel production, imposing carbon taxes, and investing in alternative sources of "green energy"—are perceived as more threatening than the cost of doing nothing about it.[32] For those who might be particularly impacted by such costs—whether we're talking about Big Oil, politicians, or blue-collar workers eking out a living in the coal industry—weighing immediate economic losses against the possibility of dooming life as we know it for future generations has the potential to create considerable cognitive dissonance. As is often the case, however, motivated denial offers a simple solution to resolve the discomfort of that calculus. Just pretend that ACC doesn't exist and—voila!—an inconvenient truth becomes a convenient untruth. But while that kind of denialism might offer short-term benefits to some, it does nothing to alter the reality of the longer-term threats that continue to loom over our heads and those of our children. While there's no doubt that denialism can be stubborn, facts can be stubborn, too.

The Hills We Die On

In February 2022, news outlets ran a story about Chad Carswell, an Air Force veteran with stage 4 renal disease who was told that he couldn't get a kidney transplant at a North Carolina hospital if he continued to refuse vaccination against COVID-19.

"I'm not getting it. There's nothing to discuss," he told the doctors.

"You know if you don't get this vaccine, you don't get a kidney and you'll die," they replied.

"Well, I guess I'll die," Carswell said. "I was born free and I'll die free."[33]

Carswell, who'd already had COVID-19 twice, explained that he wasn't "anti-vax" or a believer in vaccine conspiracy theories and he claimed that it had "nothing to do with politics." Instead, his insistence on not getting vaccinated, even though he might die over it, was "about standing up for our rights and understanding that we have a choice."[34] However, despite *The Epoch Times* running a headline claiming Carswell was "forced to choose between getting [the] vaccine or dying," he was hardly choosing to die—a month later, Carswell was working hard to get on the transplant list at another hospital in Texas.[35] In 2023, he managed to get the transplant at Duke University Hospital with a new kidney donated by his mother.

Carswell's story reminds me of an encounter I once had with one of my own patients who was hospitalized for psychiatric reasons but was also suffering from urinary retention so he that was catheterizing himself several times a day. He didn't really need to do that: since his retention was caused by an enlargement of his prostate, he'd probably be able to urinate without difficulty if he just agreed to take the medications he was prescribed. But he wouldn't do that. And he wouldn't agree to get basic labs—like a urinalysis—done either, even when he started complaining of blood in his urine. When I asked him why, he said it was because of his "religion." When I then inquired about what religion he followed or if he could explain the specifics of why it would prohibit running lab tests on the urine he'd already collected from catheterizing himself, he resisted providing any details. With some additional back-and-forth however, it became clear that when he spoke about his religion, he was just referring to his own personal and unbudging beliefs and preferences.

Values, Morals, and Identity

When it comes down to it, so much of the cognitive machinery of false belief discussed throughout this book—whether we're talking about delusions, positive illusions, fallible memories, confirmation bias, motivated reasoning, or cognitive dissonance—operates in the service of preserving a stable sense of who we are. City

University of New York philosopher Eric Mandelbaum therefore refers to this machinery as a "psychological immune system"—just like Dan Kahan's "identity protective cognition" and Jay Van Bavel's "identity-based model of belief" that I mentioned in Chapter 4—that insulates the beliefs that we've chosen to make part of our core identities from outside threats.[36] And yet, unlike our actual immune system, this isn't a healthy defense mechanism—it's one that perpetuates false beliefs and can often result in significant personal distress and social dysfunction. It isn't particularly healthy to resist changing our beliefs as if their immutability is a measure of our own integrity, just as it isn't particularly healthy to equate our beliefs with who we are.

Ever since Chapter 1, I've been making the case for instead thinking about our beliefs as probability judgments based on available evidence and understanding faith for what it is—an active choice of believing in something even though the evidence to support it is lacking. The mentally healthy way to hold beliefs and keep faith is to do so with "cognitive flexibility"—keeping in mind the possibility that we could be wrong, not insisting that we're right when we don't have the facts on our side, and being able to understand and respect other people's faith when it's just as unsupported by evidence as our own. Although "standing up for what we believe in" is often heralded as a virtue, if we keep in mind the paradox of faith, whereby we tend to feel most passionately about our beliefs when they're least supported by evidence, we might come to realize the merits of "standing down from our beliefs" and not allowing ourselves to feel threatened by changing them when we look at the evidence and listen to other perspectives.

We would also do well to ask ourselves, what hills of belief are really worth dying on? Is it really worth dying—or killing other people—over our beliefs about God and what He, She, or It might look like? Is it worth betting our lives on who's going to win the Superbowl or the World Cup? What about the beliefs that the moon landing was faked, or that vaccines cause autism, or that ACC is a hoax even though experts assure us otherwise? If the implication that few such beliefs—or even *any* beliefs—are worth dying for seems wrongheaded, it's probably because we often conflate factual beliefs with "values" and "morals."

To help us understand what values are, acceptance and commitment therapy (ACT)—a type of psychotherapy that I mentioned back in the Preface—uses a specific exercise that involves composing the epitaph on our future gravestone.[37] What, the exercise asks, would we want written there? Would we list the steadfast beliefs that we took with us—or denied—to the grave? Would "Mad Mike" Hughes, who I described in Chapter 4, have wanted his epitaph to read, "Here lies Michael Hughes: Gave his life to prove the Earth is flat"? I doubt it. And it seems just as unlikely that any of the millions who have died from COVID-19 would have wanted theirs to say "never did believe that masks or vaccines worked" or that anyone would want the cenotaph for modern civilization to read, "climate change denialists to the very end."

If Carswell's declaration that "I was born free and I'll die free"—echoing Founding Father Patrick Henry's "give me liberty, or give me death"—seems much more epitaph-worthy, it's because it isn't expressing a factual belief so much as it's articulating a value. When we talk about factual beliefs, we're talking about what we think is true, but when we talk about values, we're talking about what ACT calls a "chosen life direction" or what we think is important in our lives. Carswell didn't defend his vaccine refusal based on the belief that vaccines don't work; he defended it by saying that what he valued more than anything was freedom of choice. New School psychologist Jeremy Ginges and University of Michigan anthropologist Scott Atran use the term "sacred values" to describe values that represent "moral imperatives."[38] Values and morals aren't quite the same though: whereas values are beliefs about what's important, morals are judgments about what's "right" and "wrong" or "good" and "bad." By "sacred values," Ginges and Atran mean moral absolutes that are sacrosanct—they're zealously guarded against change because they're "intimately bound up with sentiments of personal and collective identity." When we think about values and morals in this way, wrapped up with our identities so that giving them up feels like an existential threat, it's easy to see why we might be willing to die over them.

Moral Relativism Versus Moral Absolutism

Morals aren't just tied to our own self-identity; they're also crucial to how we see others. Across several studies, University of Pennsylvania psychologist Nina Strohminger and her colleagues have demonstrated that how we perceive other people's identities is mostly determined by our perception of their "moral character," as defined by the relative presence or absence of universally valued qualities like honesty, trustworthiness, loyalty, compassion, fairness, generosity, integrity, and humility as well as their stances on widely shared morals such as murder being wrong.[39] This has been found for children and adults in studies involving appraisals of both hypothetical scenarios, as well as judgments about people affected by dementing illnesses like Alzheimer's disease. For example, perceived changes in the moral character of family members with dementia are more predictive of perceiving their essential personhood—our sense of who someone is—as changed than is the loss of other characteristics like memory, intelligence, humor, sociability, and ebullience.[40] Such studies provide support for the "essential moral self hypothesis" that tells us that, more than anything else, we define people and assess our desire to affiliate with them based on our moral perception of them as "good" or "bad."

Although psychologists, sociologists, and anthropologists generally agree that moral judgment likely evolved as a form of social cognition that allowed human beings to live communally, New York University psychologist Jonathan Haidt argues that many of our modern social conflicts stem from disagreements over moral priorities. According to "moral foundations theory" developed by Haidt and his colleagues, there are five core moral poles including care–harm, fairness–cheating, loyalty–betrayal, authority–subversion, and sanctity–degradation that in turn motivate pro-social behaviors and virtues like nurturance and compassion, reciprocal altruism and justice, coalition formation, the construction of social hierarchies, and the establishment of taboos.[41] While moral foundations theory was originally created to study cross-cultural differences, over the past decade Haidt and his colleagues have been making the case that these moral foundations provide a telling account of political differences based on evidence across several studies that conservatives organize their moral judgments around all five domains, whereas liberals tend to prioritize care/harm and fairness/cheating over the others.[42] They further argue that this partisan difference can be understood as representing two distinct strategies to suppress selfishness and promote communal living in the form of "individualizing" approaches that focus on equal rights and "binding" approaches that focus on strengthening group cohesion and identity that can come into conflict within multicultural societies. For example, moral foundations theory says that clashes over issues that lie at the heart of political division in America, like racial equity that I described in the previous chapter, can be understood as moral conflicts between the Left's prioritization of individual care and fairness versus the Right's dedication to preserving traditional institutions and ingroup identity.

Since Haidt models moral judgments as innate, intuitive, and automatic based on the kind of fast thinking that Daniel Kahneman proposed while also being intertwined with gut feelings or visceral emotion, moral foundations theory claims that affective polarization in politics is a consequence of viewing ourselves as operating from a moral high ground so that we hash out conflicts with self-righteous indignation, contempt, repugnance, and disgust. Haidt and his colleagues therefore caricature American political polarization as follows:

> liberal and conservative eyes seem to be tuned to different wavelengths of immorality. For conservatives, liberals have an "anything goes" morality that says everything should be permitted for the sake of inclusion and diversity, no matter how bizarre or depraved. For liberals, conservatives lack basic moral compassion, especially for oppressed groups, and take a perverse joy in seeing the rich get richer while innocents suffer in poverty.[43]

For his own part, Haidt claims that once he came to see political differences in this light, he "lost his righteous passion"[44] as a liberal and a Democrat so that he now identifies as a "heterodox" who views both the racism of the Right and the focus on "victimhood and victimization" by the Left as unhealthy extremes that are pushing America's political divide further apart.[45]

Although moral foundations theory has been a major influence in psychology as well as in other fields of study, it has also taken a lot of heat from critics. Some claim that Haidt's selection of five moral foundations was haphazard and not based on any underlying theory of evolved cooperation within "non–zero sum games" where both sides can win.[46] Others note more specifically that conservatives are less concerned about care and fairness than liberals and argue that the "binding" foundations that they instead favor aren't really motivated by genuine moral reasoning so much as they're driven by the kind of epistemic, existential, and system-justifying goals that I discussed in the previous chapter.[47]

The way I see it, just like using beliefs to define our identities—whether our own or those of other people—using values and morals to define them has the potential to cause significant problems within our social interactions, especially as we move beyond thinking about individuals to thinking about groups and cultures. On the one hand, shared values and morals are often what bring people together in the first place and over time they continue to serve as a rallying point for group cohesion. Some universally accepted morals—like the wrongness of murder and theft—are essential pillars that support rule of law. But, on the other hand, the universality of other moral judgments isn't as consistent so that insisting that all morals are immutable absolutes can often prove more divisive than unifying. Moral foundations theory suggests that it's the mismatch between the reality of moral relativism and the subjective perception of moral absolutism that lies at the core of political division today, just as it's differences in moral judgments that more than anything else govern partisan motivated reasoning and denialism.

It doesn't have to be this way. We would do well to understand values and morals as cognitive judgments that are subjective and relative so that, unlike factual beliefs, they can't be as easily settled with evidence. What one person or group values and moralizes, another does not—moral judgments related to democratic representation, women's rights, suicide, and even slavery and pedophilia vary across cultures and have changed considerably over time. To be clear, I'm in no way suggesting that slavery or pedophilia shouldn't be moral taboos enforced by law, but I am acknowledging that slavery was once accepted in the United States and still persists in one form or another throughout the world today just as pedophilia was culturally sanctioned in ancient Rome and medieval Japan and remains a cultural practice among the Etoro people of modern-day Papua New Guinea. If we allow ourselves to step away from the black-and-white binary bias of moral absolutism and think of moral dilemmas as non–zero-sum situations, we might be able to give up our insistence that we're always situated on a moral high ground.

In psychology and philosophy, the complexity of moral dilemmas is well-demonstrated by the "trolley problem," which poses a hypothetical scenario in which a trolley or train is on a collision course with a group of people standing on the track who can't move. A bystander observing the scene has access to a lever that can divert the trolley from its course onto another track where there's only a single person in harm's way, leaving them with a moral choice between letting five people die or killing one. When faced with this moral quandary in the psychology laboratory, an overwhelming majority opt to sacrifice one in the name of saving five. But when the scenario is modified to specify that the one person is a relative, spouse, or child, that proportion drops significantly.[48]

Although democracies are built upon universal moral foundations like equal rights and rule of law, conflict is unavoidable over some moral dilemmas akin to the trolley problem, where there isn't always a morally superior answer. Just so, although many of our most divisive political issues here in the United States—like abortion, gun control, racial equality, and proposed mandates about vaccination or curbing fossil fuel consumption—are often framed as disputes over factual beliefs or moral absolutes, they would be better understood, framed, and debated as disagreements over legislating solutions to ambiguous moral dilemmas. Viewed through this lens, the abortion debate becomes less a question of when life begins and more a conflict over the relative moral superiority of protecting the right of a woman over her own body versus the need for government to restrict that right in the name of protecting a fetus. In much the same way, gun control is less a question of whether gun ownership makes individuals safer and more a conflict over the relative merit of ensuring the unalienable right to defend oneself with a gun at all costs versus the need for government to limit that right to safeguard society against the dangers of unbridled firearm violence. Racial politics is less about how much discrimination exists in the world and more about the relative moral value of preserving the notion of equality by treating everyone the same versus promoting equity by acknowledging that an uneven playing field warrants leveling. Fighting about wearing masks and getting vaccinated during a pandemic is less about how much those interventions reduce the spread of disease and more about the comparative virtue of exercising one's personal freedom versus tolerating inconvenience in the name of altruism. And finally, climate change is less about whether global warming is anthropogenic and more about weighing the immediate economic toll of curbing fossil fuel consumption against doing everything possible to preserve living conditions on the planet for future generations.

So long as we continue to frame these issues as one-sided moral absolutes or disputes over facts and alternative facts, we remain stuck with unresolvable conflicts over zero-sum solutions. If we were to instead acknowledge and discuss them as the moral quandaries that they really are, consensus and non–zero-sum solutions might lie within our grasp.

The Five Stages of Ideological Commitment

Most of this book has been devoted to explaining why we believe falsehoods that can be refuted by evidence while steering clear of values, morals, and faiths that involve subjective "belief in" as opposed to objective "belief that," as distinguished back in Chapter 1. However, separating "belief in" and "believe that" becomes all but impossible when we start talking about "ideologies" defined as systematized sets of both factual beliefs as well as the morals and values that make up one's worldview. While ideologies can be secular or religious in their focus, religious and political scholar Shahi Hamid argues that as religious faith declines within a society, secular ideologies are increasingly supported by a kind of religious fervor so that we can think of "all deeply felt conviction [as] sublimated religion."

> Abraham Kuyper, a theologian who served as the prime minister of the Netherlands at the dawn of the 20th century, when the nation was in the early throes of secularization, argued that all strongly held ideologies were effectively faith-based,

and that no human being could survive long without some ultimate loyalty. If that loyalty didn't derive from traditional religion, it would find expression through secular commitments, such as nationalism, socialism, or liberalism. The political theorist Samuel Goldman calls this "the law of the conservation of religion": In any given society, there is a relatively constant and finite supply of religious conviction. What varies is how and where it is expressed.[49]

University of Cambridge psychologist Leor Zmigrod further argues that there are two key components of ideologies: a moral component containing "descriptive and prescriptive attitudes about social relations and norms" and a relational component that includes a "strong in-group favoritism towards adherents of the ideology coupled with distrust [of] out-groups."[50] Taking Hamid and Zmigrod's perspectives together with those of Kalmoe and Mason that I described in Chapter 8, we can therefore think of ideologies as the identity-defining intersection of individual and shared group belief that often includes "strong identification" with an in-group and "moral disengagement" with an out-group that has the potential to result in violent conflict.

Back in Chapter 1, I made the case that belief conviction is often best understood in quantitative terms. As Hamid suggests when he talks about religious fervor, that's just as true for values, morals, and ideologies if not more so. At the same time, biases, beliefs, and behavior are separable. As I argued in the previous chapter, biases don't necessarily govern beliefs, and beliefs don't necessarily dictate behavior—we can harbor unconscious implicit biases and not have conscious racist beliefs just as we can believe very strongly in ACC without doing anything about it. With Zmigrod's "relational component" in mind, the depth of our ideological conviction and whether we act upon it is often based on the extent of our affiliation with an ideological group. For these reasons, I find it helpful to quantify ideological commitment based on belief conviction, group affiliation, emotional investment, and behavioral dedication within a simple model made up of five stages of ideological commitment including "nonbelievers," "fence-sitters," "true believers," "activists," and "apostates."[51] While the model draws from research supporting an "ascending typology" of belief in conspiracy theories along with other research on the "radicalization" of those who join so-called extremist groups and cults,[52] it can be used to characterize commitment to any ideological cause. Although it's not a clinical scale—none of the prototypical examples that follow is meant to describe psychopathology per se—the model is intended to illustrate how deepening ideological commitment increases the risk of associated psychological distress, social conflict, and potential violence.[53]

At the beginning of this book, I quoted Robert Pirsig as saying that "no one is fanatically shouting that the sun is going to rise tomorrow." In a similar way, the main utility of the five-stage model of ideological commitment lies in characterizing dedication to ideologies that either diverge from the cultural mainstream or represent contested ideologies—like "pro-choice" versus "pro-life"—where there's no clear cultural default. While broadly accepted ideologies like "representative democracy" in the United States are often taken for granted and don't require any significant commitment to maintain, divergent ideologies like "communism," "authoritarianism," or "anarchy" would be expected to give rise to more cognitive dissonance and therefore require more belief conviction, emotional investment, and dedication to the cause for them to be embraced. That's what the five-stages of ideological commitment attempt to capture.

Nonbelievers

Bill is a middle-aged married man who describes himself as "middle of the road" and doesn't identify as a Democrat or Republican. He tends to endorse liberal perspectives on social issues and conservative economic stances and thinks that ranked-choice voting is worth adopting more broadly. He supports free market capitalism, but thinks that American politics needs serious campaign finance reform. He believes in the right to bear arms, owns guns for self-protection, and enjoys recreational shooting, but he favors stricter gun control laws and doesn't like what he sees as the fear-promoting propaganda of the National Rifle Association. He was raised in a Christian family, but now considers himself an agnostic. He's "pro-science," "pro-vaccine," and wore a mask in public throughout the first three years of the COVID-19 pandemic, mostly because he believed it might protect other people like his elderly parents. He belongs to few organizations, isn't particularly politically active beyond voting—he's not the type to attend rallies or marches—and doesn't consider himself much of a "joiner," although he's a big fan of the Pittsburg Steelers and "bleeds black and gold."

The term "nonbeliever" refers to someone who doesn't endorse a particular ideology in question, although that shouldn't be conflated with being a nihilist who has no beliefs, values, or morals at all.[54] Nonbelievers like Bill have opinions and ideological beliefs just the same, but they tend to acknowledge moral relativism, embrace tolerance for divergent perspectives, and avoid the "paradox of faith" that I mentioned in Chapter 1 in favor of skepticism, whereby factual beliefs are based on objective evidence. They tend to rely on analytical thinking over intuition when forming beliefs while maintaining them with cognitive flexibility. I'll say more about this mentally healthy way of believing in the next and final chapter.

Fence-Sitters

Shannon is a mother of two with a third on the way who's interested in "wellness," loves yoga, and maintains an organic, non-GMO, and gluten-free diet. In recent years, she has grown increasingly concerned about vaccines based on what she's heard from other women in her yoga class and online parenting groups on Facebook. She considers herself "pro-safety" rather than "anti-vaccine," but she's "looking for answers" and values "doing her own research." She's aware that the mainstream medical community claims that vaccines are safe, but she's watched YouTube videos by doctors like Joseph Mercola, is disturbed by what she's seen from the Vaccine Adverse Event Reporting System, and has a friend whose daughter was diagnosed with autism after receiving the vaccine for measles, mumps, and rubella. Although her first two children were fully vaccinated, she wonders whether natural immunity is a better alternative and is seriously considering not vaccinating her third child.

As I mentioned in Chapter 4, the term "fence-sitter" is often used to describe the vaccine hesitancy of people like Shannon, but it can be applied to anyone who's skeptical and mistrustful of mainstream cultural ideologies so that they're considering other counter-cultural narratives. Fence-sitters aren't yet dedicated to an ideological cause, but they're heading in that direction as they shift their epistemic affiliations—that is, who they trust as informational sources. Wrestling with the cognitive dissonance of two opposing ideologies, the more time and energy they invest in investigating a new ideological group, the stronger their ideological commitment often becomes. At the same time, fence-sitters are more likely than "true believers" to drift or be drawn back to mainstream narratives and ideologies.

True Believers

Darryl is a 35-year-old Army veteran who saw heavy combat and witnessed the death of many of his fellow soldiers and closest friends in Iraq. After falling into harmful alcohol and OxyContin use and becoming homeless, he was able to get sober after he "accepted Jesus Christ as his personal savior" and was "born again" about six months ago. Most of his social activities are now with friends from Alcoholics Anonymous (AA) or his Evangelical church. He has a girlfriend, but has taken a vow not to have sex again until he marries. With a new lease on life through his newfound calling, he's excited about becoming an AA sponsor and "spreading the Word" through an upcoming missionary trip in Central America. When an old Army buddy recently met up with him and told him that it was strange to see him talking so much about Jesus and not drinking anymore, Daryl shook his head and said, "I've been saved, brother ... I've seen the light."

Adapting a term from author and philosopher Eric Hoffer's classic treatise on mass movements,[55] "true believers" have strong ideological convictions—often in the form of sacred values—that have become core components of their identity. Darryl doesn't just believe *in* Jesus Christ as his personal savior, he strongly identifies *as* a born-again Christian. With identity and ideology so-fused, true believers often feel enlivened by having found greater meaning beyond their own lives through an ideological cause as well as a new sense of belonging from joining a subcultural group or "family." This can be mentally healthy, as in Darryl's case, but when conflicts arise between the subcultural ideologies or mainstream ideologies—typically due to behaviors that violate social norms rather than beliefs per se—they're often managed by doubling-down on ideological conviction and circling the wagons by strengthening ties to an ideological group. Perceived attacks on one's ideological beliefs can then be regarded as personal assaults and existential threats to both the individual and the group. When that happens, true believers often feel the need to take action to defend themselves.

Activists

Poe is a 22-year-old woman who has always loved animals. She became a vegetarian at age 5 and a vegan by age 7. As a teenager, she volunteered for animal rescue organizations and joined People for the Ethical Treatment of Animals. In high school, she organized protests against animal cruelty and the treatment of livestock in the farming industry. While in college majoring in agricultural science, she was inspired after learning about the Animal Liberation Front and grew increasingly alarmed about the biomedical research being conducted with primates at her university. When rallies and demonstrations calling for the research to be shut down failed to make a difference, Poe and other members of her campus animal rights group decided that they had to step up their efforts. One night, they broke into a campus laboratory, defaced and destroyed property, and attempted to free the research animals. They were arrested and brought up on charges in violation of the Animal Enterprise Terrorism Act.

Unlike true believers, it's not enough for "activists" to merely believe—they find it necessary to act upon their ideological beliefs in order to stand up for, advance, or defend a cause. Their beliefs aren't only imperative and absolute, they're also moral obligations that must be adopted by or imposed on others. Acknowledging that activism can be either healthy and productive or unhealthy and dangerous, the neutral term "activist" encompasses a wide spectrum of behavior while avoiding the subjectivity and moral relativism of terms like "extremist," "radical," and "terrorist." After all, as the saying goes, "one man's terrorist is another man's freedom fighter."

Most of us have the potential to become activists—and even violent activists—under the right conditions. Just ask yourself what it would take for you to vote in an election or donate money to a cause—hopefully not much. What about attending a rally or running for political office—probably a strong desire to support a cause and make a change? If resorting to violence in defense of a cause seems like a much bigger leap, it's probably not that hard to imagine acting violently if we were convinced that we, our family, or our country were in danger from a tangible threat. Among individuals and social groups whose ideologies and identities have merged, such threats don't have to be physical—they can just as easily be perceived as existential threats when they involve a cherished belief, principle, value, moral, or a way of life.

In the wake of events like 9/11 here in the United States and suicide bombings abroad, researchers have devoted considerable effort to determine what conditions might drive people to engage in ideologically driven violence and acts of terrorism. But, despite those efforts, no single "profile" has emerged as a reliable predictor. While ideological passion; group identification; a quest for significance; identity or existential threat; a sense of grievance, unfairness, victimhood, urgency, having exhausted alternatives, and a foreshortened future; and responses to "calls to action" are often relevant,[52,56] the myriad pathways converging on violent activism have been described in terrorism studies research as "equifinal."[57] In other words, when it comes to violent activism, there is no "one size fits all." In Poe's case, her path to ideological violence began with zealous ideological fervor for the biocentric belief that "animals are people"—a sacred value that brought her in opposition with the cultural mainstream. Although her activism was initially nonviolent and gave her a healthy sense of purpose and significance, after joining various activist groups, it became easier to engage in black-and-white moralizing that drew a bright line between the "good" ideological "family" to which she belonged and an "evil" oppressor. Once she began to view that family and the animals they were defending as aggrieved victims wronged by injustice and facing an existential threat, Poe's life took on greater significance through a "call to arms" that included resorting the violence and the possibility of sacrificing herself for a higher cause.

Apostates

There's no timelier prototype of an "apostate" than the real-life case of Jitarth Jadeja, an Australian man who fell down the QAnon conspiracy rabbit hole after listening to Alex Jones and following Infowars. With "a lot of time on his hands" while trying to complete a graduate degree, he spent two and a half years immersed in the online world of QAnon[58] while anxiously awaiting each new "Q drop"[58] so that it became "all he wanted to talk about," to the exclusion of maintaining relationships with friends.[59] Soon, Jadeja accepted QAnon dogma to the point of believing that CNN reporter Anderson Cooper was part of a Satan-worshipping cabal that ate babies[60] and coming to feel that he was part of "an existential battle between good and evil" such that "if Hillary Clinton had been executed publicly, [he] would have cheered."[59] Over time, however, Jadeja began to take note of more and more inconsistencies within QAnon theories and their failed predictions. The straw that broke the camel's back finally came when an online QAnon follower supposedly asked Q to get President Trump to say the phrase "tippy top"—when Trump did just that, Jadeja initially took it as confirmation that "Q existed and had the ear of the president."[60] But when he later found a YouTube video that debunked that conclusion with numerous examples of Trump using that phrase previously, he suddenly "realized [QAnon] was all a very slick con."[58] By 2021, Jadeja described feeling "angry and betrayed" by the movement and had found a renewed sense of purpose through his role as a senior moderator of the QAnonCasualty forum on Reddit, an online support group for family members of QAnon true believers.[61]

Just as there are many paths to becoming a true believer or an activist, there are many different paths to becoming an apostate who renounces an ideology. For Jadeja, objective evidence that gave the lie to QAnon dogma stretched his cognitive dissonance to its breaking point and brought his ideological world crashing down. In much the same way, disillusionment with a cause and "loss of faith" is the most common reason that people leave so-called cults along with having conflicts with other members of an ideological group and coming to see imperfections and contradictions within a group's leadership.[62] But Jadeja also described feeling "crushed" when he realized that QAnon was a lie—only by finding new sources of social support after writing about his experience did he start to feel "put back together again."[60] It should therefore come as no surprise that, based on research with extremist and terrorist groups, "deradicalizing" true believers and activists by attempting to "deprogram" them is often only marginally useful while "disengagement" strategies including legal consequences that separate members from an ideological group, provide contact with other apostates, and open doors to societal reintegration have a much better chance of success.[63] In other words, if we ever expect true believers and activists to become apostates, we can't just expect them to walk away by trying to reason with them over their beliefs and change their minds—leaving an ideological group often requires replacing one's ideological devotion with a healthier substitute. For Jadeja, whatever psychological or social needs might have driven him down the QAnon rabbit hole were rechanneled into helping other people find their way out like he did.

* * *

When Nobel Prize winner Adam Riess modified his beliefs about the universe to fit new data, he was excited to have made a breakthrough discovery that brought his understanding of reality closer to the truth. Giving up our beliefs admittedly gets harder when we consider the range of epistemic, existential, and relational needs that are satisfied by our ideological beliefs, but doing so doesn't have to be so threatening. In my career as a psychiatrist, I can't remember a single instance of a patient giving up their false beliefs—even when they amounted to grandiose delusions of wealth or intellect—and feeling they were worse off as a result. Embracing truth can be healing that way.

An anonymous quotation attributed to various sources through the years tells us that "we see things not as they are, but as we are."[64] But we are not our beliefs—we don't have to let our beliefs become our identities, we don't have to let disagreements be taken as personal attacks, and we don't have to allow ourselves to think of changing our minds and walking back our ideological commitments as the death of the self. Although we might feel as if we're enduring and immutable, our true selves are constantly changing over time. With that in mind, we can modify our beliefs in light of evidence—and even shift our ideologies and ideological affiliations by adopting other perspectives—while still remaining true to who we are. And when we change our factual beliefs to align with reality and moderate our ideological beliefs in ways that help us live with each other in greater harmony, we can not only remain the same person at our core, we can also grow, become better versions of ourselves, and lead happier and more meaningful lives.

10
A Prescription for a Post-Truth World

> The aim of argument, or discussion, should not be victory, but progress.
> —Joseph Joubert

> And you will know the truth, and the truth will set you free.
> —John 8:32

From Diagnosis to Cure

There's a running joke in medicine that says that psychiatrists don't really know what causes mental disorders and that on those rare occasions when an underlying biological mechanism is discovered—as in the case of neurosyphilis—they cease to be psychiatric syndromes and instead become neurological diseases. A related critique claims that while neurologists are often good at "finding the lesion" responsible for neurological disorders—as in the case of stroke—there's often not much they can offer in the way of treatment.

Within psychiatry and psychology, there's a similar stereotype that's sometimes used to argue that while "insight-oriented" talk therapies like psychoanalysis or psychodynamic psychotherapy might help us identify and understand the origins of long-standing patterns of dysfunction, they don't necessarily help us change our behavior. In contrast, cognitive behavioral therapy, as a described in Chapter 1, is more focused on modifying the underlying cognitive distortions and false beliefs that lie at the root of our dysfunctional feelings and actions, thereby resulting in meaningful change.

While none of these caricatures is particularly accurate, they nonetheless highlight the potential disconnect between determining the cause of what ails us and implementing its remedy. In this book thus far, per the stereotype of a neurologist or an insight-oriented therapist, I have mostly focused on identifying the psychological mechanisms of false belief—the cognitive pitfalls or "lesions" that lead to self-deception—with only the occasional commentary on how we might sidestep them. This final chapter therefore concludes by offering a more cohesive antidote to false belief as a kind of a prescription for a post-truth world. This prescription offers as much prevention as it does a cure, starting with recommendations for what we can do as individuals to stand down from the unwarranted overconfidence that we often have in our beliefs so that we can better align ourselves with the facts as supported by objective evidence. It then follows with recommendations for what we can do collectively as a society to promote truth as a communal value, push back against the unrelenting tide of misinformation while still protecting free speech, and hold people accountable for disinformation that harms. Finally, it concludes with advice on how we can interact with those who disagree with us by dialing down the conflict in favor of more productive engagement and dialogue.

The Holy Trinity of Truth Detection

Echoing findings that have recurred throughout these pages, the individual remedy against false beliefs can be summed up by what I call—if somewhat irreverently—the "Holy Trinity of Truth Detection" consisting of intellectual humility, cognitive flexibility, and analytical thinking. Across multiple studies, these three foundational pillars have been found to be consistently associated with a lower risk of "epistemically suspect beliefs" including belief in misinformation, fake news, anti-science rhetoric, conspiracy theories, pseudo-profound bullshit, and moral absolutism.[1]

Intellectual Humility

Simply put, intellectual humility means acknowledging that we might be wrong.[2] With this possibility in mind, we can hold beliefs as probability judgments based on available evidence rather than "truthiness" or how much we want them to be true. As I described back in Chapter 2 and in the previous chapter with the example of Adam Riess, experts who manage to avoid the unwarranted overconfidence of the Dunning-Kruger effect embody intellectual humility by acknowledging both *what* they don't know and *that* they don't know. If knowledgeable experts can say "I don't know" without allowing that admission to threaten their identity or self-worth, then so can all of us.

A good way to begin cultivating intellectual humility is to start by asking ourselves a provocative question posed in a paper by Bowling Green State University philosophy professor Hrishikesh Joshi: "What are the chances you're right about everything?"[3] Naturally, the only reasonable answer is "0%," such that only the most arrogant or delusional among us would dare claim otherwise. A related question also worth asking is "How much of what we believe is true?" Acknowledging that at least some of what we believe is probably wrong sets us on the path to intellectual humility and opens us up to learning something new.

The potential payoff of intellectual humility is well-illustrated by the phrase, "the truth comes knocking" based on a Buddhist parable about a man whose village is ransacked by bandits and burned to the ground while he's away on a trip. When he returns, he finds the village destroyed with everyone dead. Although he assumes the worst about his family, it turns out that his son was kidnapped by the bandits and survived. Years later, the son escapes, finds his way back to his father, and knocks on his door, but the father berates the boy to go away, telling him that his son has long since died. The boy begs and pleads to be let in, but after being repeatedly spurned, eventually gives up, walks away heavy-hearted, and never sees his father again. The lesson, according to the Buddha, is that "if you cling to an idea as the unalterable truth, then when the truth does come in person and knock on your door, you won't be able to open the door and accept it."[4] If we can't acknowledge the possibility that we might be wrong, we risk closing ourselves off from the truth.

Intellectual humility acknowledges that we're all vulnerable to cognitive biases, the Dunning-Kruger effect, confirmation bias, motivated reasoning and denial, susceptibility to misinformation, bullshit receptivity, cognitive dissonance, and the other various pitfalls of false belief. When we fail to see these foibles in ourselves and only recognize them in our ideological opposites, our ability to see the truth is obscured by the "bias blindspot" that I referenced in Chapter 3. Such finger-pointing is intellectual arrogance—the antithesis of intellectual humility. As I concluded in the Preface, the problem of false belief isn't one of "them" or "it," but of "us" and the world we live in. Nobody—not you, not me, not experts, not anyone—is completely immune to it.

Cognitive Flexibility

As I mentioned in the previous chapter, cognitive flexibility involves being open to changing our minds in light of evidence or upon hearing other perspectives. It means that even if we don't agree with those whose viewpoints differ from our own, we're at least willing to learn and understand where they're coming from. Cognitive flexibility allows us to recognize the paradox of faith, acknowledging that the beliefs we feel most passionately about are often the ones most likely to be wrong or to have an underlying level of complexity belied by polarized extremes. It enables us to appreciate values and morals as ambiguities depending on time and circumstance rather than being absolute truths. For example, although murder is a universal taboo exemplified by the Commandment "thou shalt not kill," killing in self-defense, war, and capital punishment are common exceptions that are often culturally sanctioned. When there's polarized disagreement over other moral judgments—like those relating to same-sex marriage, abortion, or racial equity—cognitive flexibility reminds us that the conflict might be due to a moral quandary akin to the trolley problem or that it might be rooted in tribal differences of opinion about whose rights deserve to be prioritized. Evolutionarily speaking, the utility of cultural mores is to promote harmonious communal living—when they don't fulfill that purpose and instead create conflict, they ought to be reevaluated.

Of course, the conundrum of both intellectual humility and cognitive flexibility is that when we disagree about facts, it's often because someone is right and someone is wrong. As I suggested in Chapter 8, neither intellectual humility nor cognitive flexibility should be taken to mean that everyone's beliefs are equally valid or that the truth always lies in the middle—they're not a defense of false equivalencies and "bothsidesism." And so, disagreements over when the universe was created shouldn't be resolved by splitting the difference between scientific claims that it all started with Big Bang some 15 billion years ago and creationists' insistence that God made it a mere 6,000 or 7,000 years ago. As I suggested in Chapter 9, we would be better served by forming our factual beliefs based on a combination of objective evidence—like carbon dating and the fossil record—along with expert consensus, rather than on ideological belief and dogma.

That said, cognitive flexibility allows us to acknowledge that there's potential value in the wisdom of the crowd. A long history of psychological experimentation has demonstrated that the accuracy of our judgments about the world can be increased by pooling estimates from different perspectives, just as recent research has shown that our individual estimates—at least when it comes to making numerical guesses—tend to be more accurate when we take into account the estimates of those who disagree with us.[5] In other words, when combined with cognitive flexibility, disagreement can sometimes get us closer to the truth. Accordingly, there's a lot to be said for stepping outside of our echo chambers and interacting with those who don't see eye to eye with us.

The opposite of cognitive flexibility is cognitive rigidity—just the kind of hardheadedness that lies at the heart of post-truth denialism and the insistence that we know better than experts and expert consensus. As I noted in the previous chapter, trusting evidence and experts doesn't mean being a "sheep"—that is, blindly accepting institutions of authority on faith or conforming with mainstream thinking for conformity's sake—so much as not allowing ourselves to fall victim to motivated reasoning and cognitive rigidity that insists "it's my way or the highway." Together, intellectual arrogance and cognitive rigidity create a perfect storm for embracing false beliefs with the sort of unwarranted overconfidence that's reflected in the Dunning-Kruger effect.

Analytical Thinking

While intellectual humility and cognitive flexibility are vital components of the Holy Trinity of Truth Detection that can keep our tendency for unwarranted overconfidence in check, analytical thinking is particularly useful for detecting misinformation. As I described in Chapter 2, analytical thinking refers to the kind of rational, deliberative, "slow thinking" proposed by Daniel Kahneman that acts in dynamic opposition to intuitive, automatic, and "fast thinking." In Chapters 4, 6, and 7, I noted that belief in misinformation and fake news, belief in conspiracy theories, and bullshit receptivity all tend to be associated with intuitive thinking or faith in intuition. Conversely, analytical thinking—or having an analytical cognitive style—protects us from such false beliefs.

In psychology research, analytical thinking is often measured by the Cognitive Reflection Test (CRT) that includes brain teasers like

A bat and a ball cost $1.10 in total. The bat costs $1.00 more than the ball. How much does the ball cost?[6]

For this question, the intuitive answer—and therefore an exceedingly common one among those who take the CRT—is that the ball costs 10 cents, but if we think about the question more carefully and check our reasoning, we can easily determine that this would bring the total cost to $1.20 so that the correct answer must be that the ball costs 5 cents while the bat costs $1.05. While framing analytical thinking through an example of a math problem might make it seem like it's a matter of cognitive ability, numeracy, or intelligence, that's not really the case. As I mentioned in Chapters 2 and 7, our vulnerability to cognitive biases and bullshit receptivity isn't a question of how smart we are, and the same is true of the tendency to rely on intuitive thinking over analytical thinking or vice versa.[7] More than anything, analytical thinking should be understood not as an immutable trait but as a learnable skill that simply requires that we pump the brakes on the kind of intuitive "jump to conclusions" style of cognition that's associated with delusions and other types of delusion-like beliefs. Once we do that, we can slow down and ask ourselves questions like "Is this true?" and "Am I right?" while thinking critically and skeptically before accepting gut intuitions and misinformation that represent what we want to believe.

Returning one last time to the "rigidity of the right" hypothesis that I discussed in Chapter 8, some researchers have found evidence that political conservatives are more likely than liberals to prioritize faith in intuition, feelings, and instincts over analytic thinking, evidence, and objective data, suggesting that such differences might explain why conservatives today might be more vulnerable to misinformation, conspiracy theories, and anti-science beliefs.[8] In a similar way, relying on intuitive thinking at the expense of analytical thinking has been shown to be associated with a greater propensity for both religious and paranormal belief.[9] Conversely, the idea that beliefs should be based on—and therefore change with—evidence is associated with political liberalism; acceptance of science; skepticism about religious, paranormal, and conspiratorial claims; and rejection of traditional moral values.[10]

Still, a 2021 study suggests that the prioritization of intuition over analytical thinking is more than just a habit of certain individuals or ideological groups. Analyzing the language contained within millions of fiction and nonfiction books and *New York Times* articles from 1850 to 2019, the study found that the use of words related to intuitive fast-thinking like "feel," "trust," "faith," "hope," and "belief" declined steadily during the 1800s through the 1970s, but has since increased sharply over just the past few decades.[11] Conversely, the use of words related to rational slow-thinking like "science," "data," "hypothesis," "analysis," "statistics," "fact," and "conclude" rose through the 1970s but has since taken a nosedive. A similar pattern was identified whereby the use of plural pronouns like "we" and "they" have been increasingly supplanted by individual pronouns like "I" and "he" since the 1970s. It therefore appears that today's post-truth world is reflected in how we as a society have replaced rational and collective language like "we conclude" with intuitive and individualistic—we might even say narcissistic—statements like "I believe."[12] This in turn suggests that the prioritization of intuition over analytical thinking and overconfidence in our personal beliefs isn't just a trend rooted in individual psychology or the motivated reasoning of a single political party but is also an emergent cultural phenomenon that has become much more universal.

Truth, Justice, and a Better Tomorrow

That the ascendancy of intuitive thinking over analytical thinking represents a relatively recent reversal of a previously long-standing cultural trend suggests that it should be possible to swing the pendulum back again. But, thinking back to the joke about the necessity of lightbulbs wanting to change from Chapter 8, I know all too well that it's not enough to have people read this book, tell them to follow the Holy Trinity of Truth Detection, and call me in the morning. If we are to become individuals, communities, nations, and a world less vulnerable to false belief, it will require a collective desire to change. Fortunately, there are several initiatives that could help make that happen at the societal level without having to put every one of us on a psychiatrist's couch.

Educational Reform

Recalling from the previous chapter that cultural values and ideals can be powerful drivers of both group belief and behavior, one way to get the pendulum swinging back toward analytical thinking is to embrace objective truth—as opposed to only personal truths—as a cultural value. In the 1940s, the phrase "Truth, Justice, and the American Way" was coined as a patriotic slogan for the Superman comic book during World War II.[13] Since then, it has been revised several times in response to ebbs and flows of American nationalism, with DC Comics last changing it to "Truth, Justice, and a Better Tomorrow" in 2021. While that might sound a little corny, most slogans are. And, corny or not, it's just the kind of new *zeitgeist* we could use right now to kickstart a campaign against the kind of post-truth politics that exploits our collective vulnerability to overconfidence, confirmation bias on steroids, motivated reasoning, and denialism so that we ignore evidence, dispute facts, dismiss expertise, and embrace misinformation.

While a slogan can help to promote principles, the vital next step is to teach them—like the Holy Trinity of Truth Detection—through education. While large-scale educational reform might sound like a pipe dream, it's hardly unprecedented here in the United States. In the 1950s, science, math, and engineering education was overhauled after the Soviet Union launched Sputnik 1—the first ever space satellite—with the hope that a new generation of Americans would win the space race and become leaders in developing new technology. With that goal since realized, what the world really needs right now isn't more scientists, technology, or people traveling into space but another retooling of education from the ground up with an emphasis on the philosophy of science as much as on its practice. We need to do a better job of teaching new and current generations how to look to the scientific method to separate fact from fiction, following the example of a new kind of superhero—like Emily Rosa from Chapter 7—that we can celebrate as champions of truth.

As a society, we must also find ways to help people become better consumers of information, taking steps to resolve our deficits in informational literacy. In Chapter 4, I noted that few of us have ever taken a course on how to search for information in the internet era—we just type a question into Google or ask Siri, Alexa, or ChatGPT and wait for the answer or we watch CNN and Fox News and soak up like a sponge what their commentators tell us. But while access to our "peripheral brains" might make us feel smarter, as I discussed in Chapter 3, a public that can recognize bias among informational sources, acknowledge confirmation bias on steroids and motivated reasoning in ourselves, and employ data reasoning skills without succumbing to denialism would be much better informed. As I mentioned in Chapter 7, college courses like "Calling Bullshit" at the University of Washington deserve to be rescaled and implemented as early as elementary school. But it's not enough to start with the next generation: we need to better equip parents, grandparents, and everyone else to become better at distinguishing reliable information from misinformation within the flea market of opinion through the kind of media awareness campaigns like "Think Before You Click" that have been implemented across the world over the past several years.[14]

In 2015, Finland launched a war on disinformation through an educational campaign that teaches media literacy and critical thinking beginning in kindergarten and continuing through grade 12 and beyond into old age. Meanwhile, Finland's president Sauli Niinistro called upon every citizen in his country to take on the fight. The results have been impressive, with Finland ranking number 1 among its European peers in metrics of both resistance to post-truth politics and overall happiness for five years in a row.[15] A 2019 study found that those participating in Finland's educational program outperformed US students in their ability to judge the reliability of media sources and evaluate the strength of evidence presented in news articles.[16] Clearly then, educational reform is possible, although some of us have a lot of ground to make up.

Content Moderation, Censureship, and Censorship

Recalling Brandolini's Law that I mentioned in Chapter 7, the most evidence-based interventions to decrease belief in misinformation aren't debunking efforts that are reactive, but proactive "prebunking" or "inoculation" strategies that warn about misinformation in advance, beating it to the punch and building immunity against false beliefs in the process.[17] While there's little doubt that such efforts must be implemented more widely if we hope to gain ground in the war on misinformation, the sheer volume of misinformation and disinformation that's out there in the world makes doing so akin to a Sisyphean game of Whack-a-Mole.

Perhaps as a result, many have instead focused on cutting misinformation off at its source, particularly within social media platforms like Facebook, Twitter, and YouTube. While the effectiveness of such efforts—including the restriction of anti-vaccine, QAnon, and other conspiracy theory content in recent years—and their potential for a so-called backfire effect that increases rather than decreases belief in misinformation is debatable,[18] there's no doubt that they've been increasingly countered with charges of censorship and governmental propagandizing in recent years. Taking such allegations seriously, US District Judge Terry Doughty ruled in 2023 that the Biden administration had infringed free speech by strong-arming social media companies to remove misinformation related to vaccines and unsubstantiated claims about the origin of COVID-19 during the pandemic in what he saw as "an almost dystopian scenario" in which the government was acting as if it was *1984*'s "Ministry of Truth."[19] Doughty therefore imposed a ban on the administration, prohibiting it from any further contact with social media platforms with the intent of removing content deemed to represent harmful misinformation. After Doughty's ruling was upheld by the US Court of Appeals for the Fifth Circuit, the Biden administration appealed it, taking the case to the highest court of the land. In 2024, the US Supreme Court overturned the ruling in a 6–3 decision.

While such cases will no doubt continue to be brought to court and adjudicated well into the future, it's worth reminding ourselves that free speech has its limits. For one thing, the Constitutional right to free speech doesn't give us carte blanche to say whatever we want to whomever we want. While the legality of metaphorically shouting "fire" in a crowded movie theater and causing a stampede has been debated in landmark cases like *Schenk v. US* in 1919 and *Brandenburg v. Ohio* in 1969, it's a fact that speech that can be shown to be defamatory, to incite violence, or to cause injury could at the very least result in criminal liability.[20] And so, not all speech is fully protected. Being able to speak freely should never be mistaken for being granted immunity to the consequences of that speech.

Furthermore, while free speech may allow us to speak our minds and proselytize on a public street corner, the First Amendment prohibition against the infringement of speech only applies to government. Indeed, the US Supreme Court has made it clear over the past decade that private social media platforms—just like newspapers and TV networks—aren't obliged to supply everyone with a pulpit and a megaphone. Content moderation that restricts, removes, or fact-checks misinformation and other material that violates a platform's rules of conduct doesn't fall under the umbrella of Constitutionally prohibited censorship.

In much the same way, when social media platforms suspend or ban the accounts of those who violate community standards, we shouldn't allow ourselves to so easily cry foul by invoking the now popular but ill-defined term, "cancel culture." While those lamenting cancel culture today often equate it with restriction of free speech and undeserved punishment in the form of withdrawal of support, boycotting, or public shaming motivated by political correctness gone wild, a 2021 Pew Poll found that those familiar with the term are far more likely to view it as an action taken to "hold others accountable" for their words and actions.[21] When users are allowed to post misinformation and inflammatory content on social media, there's no requirement that we have to listen, just as there's no restriction against responding in kind. Like it or not, the ability of an audience to "call out" and castigate those who say things that we don't want to hear *is* a form of protected free speech. Censureship isn't censorship.

It should be noted that while most in the United States today do have concerns about the restriction of free speech in today's polarized political climate,[22] content moderation on social media has substantial public support. A 2023 Pew poll found that a majority of US adults believed that the government (55%) and tech companies (65%) "should take steps to restrict false information online, even if it limits freedom of information," up from 39% and 56%, respectively, in 2018.[23] However, a partisan divide was evident in the results, with only 39% of self-identified Republican-leaning respondents supporting such governmental restrictions compared to 70% of Democrat-leaning respondents. These findings match those of another recent study that found that 58–71% of US respondents would prefer to remove social media posts about climate change denial, anti-vaccine misinformation, election denial, and Holocaust denial with a similar partisan split.[24]

In light of what we know about partisan motivated reasoning, it should come as no surprise that we're more likely to support unbridled free speech when we're invested in protecting statements that align with our own worldviews and moral judgments, whereas we're more willing to restrict it when we're trying to quash what we find objectionable.[25] For example, while most in the United States generally oppose book bans, many liberals have supported certain books containing racist tropes or using the "N word" getting the axe in recent years, while conservatives have been proposing and passing legislation banning material that includes sexual content or LGBTQ themes. Likewise, although it's mostly conservatives who are railing against a so-called censorship industrial complex right now,[26] it was mostly liberals that were pushing for the First Amendment to apply to private businesses during the first half of the twentieth century.[27]

Such inconsistency highlights that we should all be concerned about the restriction of free speech—especially by the government—since the scope of that restriction could very well change in ways that we would find objectionable going forward. The framers of the Constitution saw fit to safeguard free speech and institute checks and balances within government because they were all too aware of the potential for a "tyranny of the majority" or a "tyranny of the masses." In a country where there's now a sharp but narrow (49–51%) political divide where compromise has become a dirty word, we now risk the tyranny of the slim majority paving a path toward authoritarian rule and moving forward with what it wants while discounting minority interests. As I discussed in Chapter 5, authoritarian regimes do indeed have a long history of stifling dissent by limiting free speech and using state-manufactured propaganda to oppress the masses by bulldozing over the truth. If we are to be spared that fate, the preservation of democracy will hinge upon our ability to instead engage in debate with our ideological opposites—whether online, within schools, in public forums, or across the Thanksgiving dinner table with friends and family—in ways that respect the rules of civil discourse while also allowing truth to prevail over false beliefs.

Vox Populi, Vox Dei

Thinking back to the trolley problem that I discussed in Chapter 9, we can see that there's a moral dilemma lying at the heart of current debates about how we should adjudicate whose opinions are right and whose are wrong. On the one hand, ceding control to the government and other institutions of authority as the ultimate arbiters of truth risks submitting to authoritarian rule. On the other, opposing content moderation that reins in misinformation risks harm from continuing to make consequential decisions based on false beliefs. Which is the greater evil: the censorship industrial complex or the disinformation industrial complex?

Rather than trying to defend one answer over another, we should recognize that there's ample room to strike a healthy balance between open discourse and the regulation of misinformation. Few of us would refute that some degree of regulation is a public health necessity to enforce the accuracy of product labeling and prohibit unsubstantiated claims about food or medications to prevent harm. We should think of misinformation more broadly in much the same way by aspiring to become a culture that demands accuracy backed up by objective evidence from those claiming not only subjective personal opinions, but objective facts. To that end, if we are to ever emerge from the post-truth world and achieve a better tomorrow, we need to establish not only truth but also justice in the form of epistemic accountability as a communal value. In the poll that I mentioned earlier in which a majority supported the removal of social media posts containing misinformation, a substantial majority also supported some degree of penalty—such as warnings, suspensions, or bans—for accounts that spread misinformation.[21] And yet, despite Elon Musk's affinity for the phrase, "*vox populi, vox dei*" (translated as "the voice of the people is the voice of God"), we've seen the opposite occur on his platform ever since he acquired Twitter in 2022 and subsequently rebranded it as X. Previously banned accounts have been reinstated; "blue checks" that once verified account identities including those of celebrities, government officials, and scientific experts have been converted into a meaningless "pay to play" commodity; and the safeguards of content moderation have been all but eliminated. Not unsurprisingly, the result has been a commensurate rise in hate speech, disinformation, and authoritarian state propaganda.[28] Only after antisemitic and neo-Nazi posts were followed by fake AI-generated nude photos of Taylor Swift going viral at the start of 2024 did Musk relent by establishing a "trust and safety center" and hiring 100 content moderators back to the platform.[29]

When falsehoods—and especially deliberate falsehoods—cause harm, we should recognize the distributed responsibility and liability for misinformation that I called for in Chapter 5. With Alex Jones and Fox News being found guilty in their respective defamation trials that I mentioned at the end of that chapter, there's now a clear legal precedent establishing potential liability for apex predators in the disinformation food chain who propagate harmful untruths. That's progress in the war on misinformation. An epistemically healthy society shouldn't tolerate trafficking in deliberate lies. It should instead hold the politicians, corporate leaders, entrepreneurs, corrupt institutions of authority, and even the social media platforms that profit from spreading disinformation accountable for spreading falsehoods rather than continuing to sweep such behavior under the post-truth rug of business as usual.

Looking to the future, another way to sidestep the moral dilemma of harmful free speech versus censorship is to focus on finding novel non-zero-sum solutions. For example, rather than putting all our eggs into the basket of content moderation, governmental oversight, and legal action against harmful misinformation, we should aspire to become a self-policing culture that's quick to call out falsehoods and bullshit for what they are while empowering consumers to weigh in on the reliability of the information that we see. That's less intellectual arrogance, censorship, or evidence of cancel culture than it is the kind of peer review that keeps academics and scientists honest as I described in Chapter 4. Indeed, recent research has shown that adding clickable "misleading" tags to posts next to the "like" button on social media sites like Twitter can reduce the sharing of misinformation online even when the content is politically polarizing.[30] Displaying a tally of misleading tags on posts was more effective than merely encouraging users to engage in analytical thinking, though there's no reason to limit ourselves to only one intervention or the other. These preliminary results suggest that *vox populi, vox dei* could be a viable guiding principle for promoting truth in public forums, although it would likely be vulnerable to echo chamber effects if "the people" are limited to a narrow, ideologically homogeneous user group. As I suggested earlier, the wisdom of the crowd is enhanced when the crowd is diverse.

Instead of focusing only on misinformation going forward, it's also worth spending more time and energy figuring out how to make truth and objectivity more appealing, competitive, and profitable so that the free market of opinion can become less of a flea market. We shouldn't resign ourselves to allowing fake news to travel faster and farther than reliable information—recalling my metaphor of thinking about information like food from Chapters 4 and 5, we should instead commit to replacing our ill-advised, child-like preference for daily candy and soda with a balanced, adult diet that includes a variety of healthy and nutritious options. Given the central role of mistrust in our vulnerability to belief in misinformation and the importance of placing the horse of trustworthiness before the cart of trust that I discussed in Chapters 5 and 6, however, this will require that institutions of epistemic authority commit to regaining the public's trust by becoming more transparent and accessible. Whether governmental, scientific, or corporate in nature, such institutions should let more of the public "in the door," so to speak, giving them a "seat at the table" where there's room to learn while witnessing and engaging in legitimate and open debate. Within academic publishing, where methodological analysis and data interpretation requires expertise, we would do well to push back against the kind of predatory journals that I described in Chapter 4 in favor of beefing up peer review and requirements for disclosing conflicts of interest. Finally, to win over "hearts and minds" in the war on misinformation, we could use not only more superheroes and role models like Chapter 8's Emily Rosa, but a new generation of nonpartisan Walter Cronkites and Ted Koppels to bring back objective news reporting as trusted "navigators" or "guides" in our search for truth.

Can't We All Just Get Along?

Any writer lucky enough to reach a large audience knows there will always be some readers who love what they're saying and others who hate it. It comes with the territory. Over the past decade since I've been writing about conspiracy theories and other delusion-like beliefs on my blog, I've certainly received my share of hate mail telling me that I don't know what I'm talking about. Readers insist that the Earth is flat. That gang-stalking is real. That COVID-19 was created in a lab as a bioweapon. That Bill Gates and Anthony Fauci really were trying to microchip people through vaccines. Or that mental disorders don't exist and that psychiatrists aren't real doctors. Meanwhile, I've been called an "idiot," a "fool," "obtuse," and much worse.

It's therefore all too easy to anticipate that some readers who actually finish this book—provided they made it past my first casual mention of the potential merits of affirmative action in Chapter 2, my excoriation of President Trump's exploitation of the illusory truth effect in Chapter 5, Chapter 8's portrayal of political division in the United States as ultimately boiling down to racial identity, and my use of the word "insurrection" across multiple chapters to characterize the events of January 6, 2021—will cry foul, accusing me of using a bunch of psychological pseudoscience to justify what amounts to a liberal screed. With my pointed critiques of anti-vaxxers, the wellness industry, QAnon, and climate change denialists, others will no doubt accuse me of being an "elite," a "Pharma shill," or a blue-pilled "sheep" who's the real victim of the bias blindspot, failing to recognize that I'm the one who's succumbing to confirmation bias, motivated reasoning, lack of critical thinking, and belief in misinformation. "I'm not biased," they'll say, "you're biased." "I'm not wrong," they'll tell me, "you're wrong."

I have two responses to such claims. First, arguments that boil down to, "I'm rubber, you're glue, whatever you say bounces off me and sticks to you" have become an all-too-common tactic from the motivated reasoning playbook these days. But, following my own advice about the Holy Trinity of Truth Detection, I concede that maybe they're right. I have my biases. We all do. And we should all get in the habit of looking in the mirror, acknowledging them, and checking them at the door, especially when we engage with those who disagree with us. In the Preface and at the start of this chapter, I wrote that the problem of false belief isn't one of "them" or "it," but "us" and the world we live in. This book isn't about finger pointing, it's about our universal vulnerability to false belief. So, when I say that all of us should look in the mirror, I really mean *all* of us.

Second, although I'm confident that the facts are on my side, we're often better off shifting the conversation away from disputes over who's right and who's wrong when we engage in dialogue with our ideological opposites, especially when one or both of us are "true believers" whose very identities are threatened by having our beliefs challenged. Engaging in the kind of debates that I complained about in Chapter 3 with two opponents insisting they're right and never backing down just doesn't get us anywhere. Although such disputes have become the status quo, whether in politics or on social media, they lack any willingness to concede that anyone might be wrong or that there could be another way of looking at things, so that there's no real exchange of ideas and no resolution. The ancient Greeks labeled this style of argumentation *eristic*—arguing as an end in itself rather than to arrive at the truth or to resolve conflict. But we can do better than that. Regardless of whether we think we're on the right side of the truth, we would do well to shed our righteous indignation about it and be more open to understanding other viewpoints and learning from each other. Instead of arguing about who's right, we can examine the evidence together and talk about how to further test a hypothesis. And we can leave aside our respective myside biases in favor of stepping back to look at the bigger picture and seeking common ground. Some academic debate clubs utilize "turncoat debating" that has participants present both sides of an argument with a concluding synthesis. In a turncoat debate, there is no opponent: a lone debater effectively argues against herself and is judged based on her ability to present two opposing viewpoints in a balanced way. We could learn a lot more from each other if we adopted that kind of opponent-less approach more often.

A few years ago, I was asked to provide a psychiatric perspective on conspiracy theory beliefs for a documentary about flat Earthers called *Behind the Curve*. While most of the footage from my interview ended up on the cutting room floor, the film nonetheless handed me my five minutes of fame after it was released on Netflix and became something of a sleeper hit. Now, I may be biased, but I think *Behind the Curve* owes part of its success to the strategy that I recommended to the filmmakers before they started shooting. Indeed, just as I advised, the movie avoids the easy trap of ridiculing flat Earthers according to the pop culture stereotype of tinfoil hat wearing basement dwellers. Instead of arguing facts, it engages with them as people looking for answers while providing a nonjudgmental response when the evidence that the flat Earthers gather fails to support their beliefs. Throughout the film, *Behind the Curve* humanizes its subjects while providing an explanation to the audience of how it's possible that relatively normal people can come to believe in something as patently false as a flat Earth and a worldwide conspiracy to conceal its flatness from the public. I've tried to model that approach in this book as best I can.

Of course, as I discussed back in Chapter 3, avoiding arguments and disparaging our ideological opposites is easier said than done, especially when we're online. But that doesn't mean that doing so lies beyond our grasp. All it really takes is a dedication to interacting with other people the same way we would if we were face to face instead of hiding behind our online anonymity and letting the road rage effect get the better of us. Taking care not to offend other people isn't giving in to political correctness that's spiraled out of control; it's simply observing the basics of common civility and the Golden Rule.

In Chapter 3, I also mentioned that my college education at MIT taught me that engineering solutions to problems is often more useful than regurgitating known facts. Nearly all of the exams I took there were "open book"—we weren't graded on our ability to memorize and recall details, but rather on how we applied those details to solve a problem. What's missing from that approach—and our lives today—is how to solve problems by collaborating with people who might disagree with us. Indeed, as I suggested in Chapter 8, we don't necessarily need to agree to get beyond ideological polarization so much as we need to stop arguing and commit to working together on communal goals.

The first physical fight I ever had with someone other than my brother occurred at summer camp when I was around seven years old. I have no memory of what it was about, but I do remember being shocked when the other kid punched me in the mouth and then throwing him to the ground, causing his knee to be bloodied. When the camp counselors broke up our little skirmish, they made us stand on either side of a tree holding hands so that our arms formed a circle around it. Eventually they said we could stop and told us to shake hands and get back to camp life. Looking back on it now, a better solution might have been to assign us to work together on some task, whether assembling the ingredients for s'mores or co-captaining a team for a game of capture the flag. That's the kind of directive that we need right now—working toward shared objectives that steer us away from ideological and affective polarization in favor of collaborating to resolve conflict and solve problems for the greater good. The more we're willing to do that in person, partnering with our neighbors and ideological opposites alike during face-to-face interactions in the spirit of compassion, mutual respect, and community instead of limiting ourselves to anonymous online exchanges and theatrics intended for an audience, the more likely we are to succeed.

"Unless"

Of all the beliefs we—and even experts—tend to hold with the most unwarranted overconfidence, predictions about the future often sit at the top of the list. And so I won't wager a guess as to whether we'll manage to forego our stubborn conviction in our false beliefs in exchange for objective truth on any meaningfully large scale. On paper, the path laid out here to reclaim Truth, Justice, and a Better Tomorrow is easy enough to follow. But while there's no doubt that we *can* do it, there's no way to know whether enough of us *will*. The only fitting conclusion of a book about false belief is to concede that I have no idea how things will turn out.

Still, harkening back to the warning I gave in the Preface that this would be more of a dark cautionary tale than an inspirational fantasy, it's certainly possible that false beliefs and self-deception might prevail over truth. The psychological cards of false belief are stacked too high against us. Not enough lightbulbs want to change. The profits made by the predators atop the disinformation food chain who benefit from the conceit that truth doesn't exist are too substantial. Too many partisans want a divorce instead of a marriage. Seeing past moral and tribal differences to appreciate diversity of belief as a communal value runs too much against our natural instincts. And far too many of those who fall victim to self-deception insist that it's not them, but their ideological opposites who wouldn't recognize a fact if it hit them in the face. It's therefore entirely possible that we'll lose the war on misinformation and succumb to false belief in the end so that previously unified societies will fracture, democracies will crumble and authoritarian regimes will arise from the wreckage, millions more will die from ignoring scientific evidence, and the planet will burn.

If we are to part ways on a somewhat more optimistic note, I'll have to borrow the ending from my favorite cautionary tale as a child, *The Lorax* by Dr. Seuss. At the end of that melancholy story documenting the environmental destruction of a mythical utopia of Truffula Trees, Swomee-Swans, Humming Fish, and Bar-ba-loots at the hands of one man's personal greed, the Lorax—the would-be defender of the land with his furry little orange body and a billowing mustache—leaves a cryptic parting word in his defeated wake: "unless." When years later, a lone remaining Truffula seed is left to the story's young protagonist so that he might replant the forests and restore the ecosystem, the meaning of the word becomes clear:

UNLESS someone like you
Cares a whole awful lot,
Nothing is going to get better.
It's not.[31]

References

Preface

1 Pierre JM. The top 5 questions everyone asks a psychiatrist. *Psych Unseen*. August 17, 2020. https://www.psychologytoday.com/us/blog/psych-unseen/202008/the-top-5-questions-everyone-asks-psychiatrist
2 Pierre JM. A mad world. *Aeon*. March 19, 2014. https://aeon.co/essays/do-psychiatrists-really-think-that-everyone-is-crazy.
3 Freud S. *The Psychopathology of Everyday Life*. W. W. Norton, 1965.
4 Wick D, Konrad C, Mangold J, et al. *Girl, Interrupted*. Columbia Pictures, 2000.
5 Mackay C. *Extraordinary Popular Delusions and the Madness of Crowds*. Richard Bentley, 1841.
6 Hayes SC, Strosahl KD, Wilson KG. *Acceptance and Commitment Therapy: An Experiential Approach to Behavioral Change*. Guilford, 1999.
7 Pierre JM. The borders of mental disorder in psychiatry and the DSM: Past, present, and future. *Journal of Psychiatric Practice* 2010;16:375–386; Pierre JM. Mental illness and mental health: Is the glass half empty or half full? *Canadian Journal of Psychiatry* 2012;57:651–658.

Chapter 1

1 American Psychiatric Association. *Diagnostic and Statistical Manual of Mental Disorders* (5th ed.). American Psychiatric Association, 2013; Pierre JM. Mental disorder vs normality: Defining the indefinable. *Bulletin of the Association for the Advancement of Philosophy and Psychiatry* 2010;17:9–11.
2 Cermolacce M, Sass L, Parnas J. What is bizarre in bizarre delusions? A critical review. *Schizophrenia Bulletin* 2010;36:667–679; Spitzer RL, First MB, Kendler KS, Stein DJ. The reliability of three definitions of bizarre delusions. *American Journal of Psychiatry* 1993;150:880–884.
3 Plato. *Phaedrus*. Penguin, 2005.
4 Pierre JM. Conspiracy theory belief: A sane response to an insane world? *Review of Philosophy and Psychology* 2023. https://doi.org/10.1007/s13164-023-00716-7
5 Kendler KS, Glazer WM, Morgenstern H. Dimensions of delusional experience. *American Journal of Psychiatry* 1983;140:466–469; Peters ER, Joseph SA, Garety PA. Measurement of delusional ideation in the normal population: Introducing the PDI (Peters et al: Delusional Inventory). *Schizophrenia Bulletin* 1999;25:553–576.
6 Pierre JM. Faith or delusion: At the crossroads of religion and psychosis. *Journal of Psychiatric Practice* 2001;7:163–172.
7 Pierre JM. Integrating non-psychiatric models of delusion-like beliefs into forensic psychiatry assessment. *Journal of the American Academy of Psychiatry and the Law* 2019;47:171–179.
8 Beck, AT. Thinking and depression. I. Idiosyncratic content and cognitive distortions. *Archives of General Psychiatry* 1963;9:324–333.

9. Burns DD. *Feeling Good: The New Mood Therapy*. Avon, 1980.
10. Pechey R, Halligan PW. Exploring the folk understanding of belief: Identifying key dimensions endorsed in the general population. *Journal of Cognition and Culture* 2012;12:81–99.
11. Cambridge Dictionary. Belief. https://dictionary.cambridge.org/us/dictionary/english/belief#
12. Schwitzgebel E, Belief. In: Zalta EN, ed. *The Stanford Encyclopedia of Philosophy (Fall 2019 Edition)*. https://plato.stanford.edu/archives/fall2019/entries/belief/.
13. McKay RT, Dennett DC. The evolution of misbelief. *Behavioral and Brain Sciences* 2009;32:493–561.
14. Simonsen A, Fusaroli R, Petersen ML, et al. Taking others into account: Combining directly experienced and indirect information in schizophrenia. *Brain* 2021;144:1603–1614.
15. Dudley R, Taylor P, Wickham, Hutton P. Psychosis, delusions and the "jumping to conclusions" reasoning bias: A systematic review and meta-analysis. *Schizophrenia Bulletin* 2016;42:652–665.
16. Verdoux H, Maurice-Tison S, Gay B, van Os J, Salamon R, Bourgeois ML. A survey of delusional ideation in primary-care patients. *Psychological Medicine* 1998;28:127–134; Pechey R, Halligan P. The prevalence of delusion-like beliefs relative to sociocultural beliefs in the general population. *Psychopathology* 2011;44:106–115; Freeman D, Pugh K, Garety P. Jumping to conclusions and paranoid ideation in the general population. *Schizophrenia Research* 2008;102:254–260; Ward T, Peters E, Jackson M, Day F, Garety PA. Data-gathering, belief flexibility, and reasoning across the psychosis continuum. *Schizophrenia Bulletin* 2018;44:126–136.
17. Freling TH, Yang Z, Saini R, Itani OS, Abualsamh RR. When poignant stories outweigh cold hard facts: A meta-analysis of the anecdotal bias. *Organizational Behavior and Human Decision Processes* 2020;160:51–67.
18. Freud S. *The Future of an Illusion*. W.W. Norton, 1961.
19. Ross L, Ward A. Naïve realism in everyday life: Implications for social conflict and misunderstanding. In: Brown T, Reed ES, Turiel E, eds. *Values and Knowledge*. Erlbaum, 1996, 103–135.
20. Pierre JM. The neuroscience of free will: Implications for psychiatry. *Psychological Medicine* 2014;44:2465–2474.
21. Seth AK. The real problem. *Aeon*. November 2, 2016. https://aeon.co/essays/the-hard-problem-of-consciousness-is-a-distraction-from-the-real-one
22. Gawande A. The mistrust of science. *The New Yorker*. June 10, 2016. https://www.newyorker.com/news/news-desk/the-mistrust-of-science
23. Hardwig J. The role of trust in knowledge. *Journal of Philosophy* 1991;88:693–708.
24. Pirsig R. *Zen and the Art of Motorcycle Maintenance*. Bantam, 1974.
25. Hebrews 11:1. King James Bible. https://www.biblegateway.com/passage/?search=Hebrews%2011%3A1&version=KJV
26. Montaigne M, Cotton C, Hazlitt WC. *The Essays of Montaigne*. Reeves and Turner, 1877.
27. Nietzsche F. The antichrist. In: Kaufman W, ed. *The Portable Nietzsche*. Viking, 1954.
28. Dennett D. *Brainstorms: Philosophical Essays on Mind and Psychology*. MIT Press, 2007; Frankish K. A matter of opinion. *Philosophical Psychology* 1998;11:423–443.
29. Price HH. Belief "in" and belief "that." *Religious Studies* 1965;1:5–27; Byrne J. Believe in or belief that? https://sites.google.com/site/skepticalmedicine/believe-in-or-believe-that
30. Geiderman JM. Faith and doubt. *JAMA* 2000;283:1661–1662.
31. Gervais WM, Norenzayan A. Analytical thinking promotes religious disbelief. *Science* 2012;336:493–496; Pennycook G, Cheyne JA, Seli P, Koehler DK, Fugelsang JA. Analytic cognitive style predicts religious and paranormal belief. *Cognition* 2012;123:335–346.

Chapter 2

1. Perry DF, DiPietro J, Costigan K. Are women carrying "basketballs" really having boys? Testing pregnancy folklore. *Birth* 1999;26:172–177.
2. Thomson GH. Should we teach statistics in the senior high school? *The Mathematics Teacher* 1924;17:129–139; Boaler J, Levitt SD. Modern high school math should be about data science—not Algebra 2. *Los Angeles Times*. October 23, 2019. https://www.latimes.com/opinion/story/2019-10-23/math-high-school-algebra-data-statistics
3. Tversky A, Kahneman D. Belief in the law of small numbers. *Psychological Bulletin* 1971;76:105–110.
4. Griffin D, Tversky A. The weighing of evidence and the determinants of confidence. *Cognitive Psychology* 1992;24:411–435.
5. Bar-Hillel M. The base-rate fallacy in probability judgments. *Acta Psychologica* 1980;44:211–233.
6. Matarazzo O, Carpentieri M, Greco C, Pizzini B. The gambler's fallacy in problem and non-problem gamblers. *Journal of Behavioral Addictions* 2019;8:754–769.
7. Tversky A, Kahneman D. Judgment and uncertainty: Heuristics and biases. *Science* 1974;185:1124–1131.
8. Kahneman D. *Thinking, Fast and Slow*. Farrar, Straus and Giroux, 2011.
9. Kvam PD, Pleskac TJ. Strength and weight: The determinants of choice and confidence. *Cognition* 2016; 152:170–180.
10. Benson B. Cognitive bias cheat sheet, simplified. *Medium*; January 7, 2017; Haselton MG, Nettle D. The paranoid optimist: An integrative evolutionary model of cognitive biases. *Personality and Social Psychology Review* 2006;10:47–66; McKay RT, Dennett DC. The evolution of misbelief. *Behavioral and Brain Sciences* 2009;32:493–561.
11. Stanovich KE, West RF. On the relative independence of thinking biases and cognitive ability. *Personality Processes and Individual Differences* 2008;94:672–695.
12. Shariatmadari D. Daniel Kahneman: "What would I eliminate if I had a magic wand? Overconfidence." *The Guardian*. July 18, 2015.
13. Taylor SE, Brown JD. Illusion and well-being: A social psychological perspective on mental health. *Psychological Bulletin* 1988;103:193–210.
14. Brown JD. Understanding the better than average effect: Motives (still matter). *Personality and Social Psychology Bulletin* 2012;38:209–219.
15. Moore DA. Not so above average after all: When people believe they are worse than average and its implications for theories of bias in social comparison. *Organizational Behavior and Human Decisions Processes* 2007;102:42–58.
16. Colvin CR, Block J, Funder DC. Overly positive self-evaluations and personality: Negative implications for mental health. *Journal of Personality and Social Psychology* 1995;68:1152–1162.
17. Anderson C, Brion S, Moore DA, Kennedy JA. A status-enhancement account of overconfidence. *Journal of Personality and Social Psychology* 2012;103:718–735.
18. Robins RW, Beer JS. Positive illusions about the self: Short-term benefits and long-term costs. *Journal of Personality and Social Psychology* 2001;80:340–352.
19. Baumeister RF. The optimal margin of illusion. *Journal of Social and Clinical Psychology* 1989;8:176–189. See also Makridakis S, Moleskis A. The costs and benefits of positive illusions. *Frontiers in Psychology* 2015;6:859.
20. Taylor SE, Brown JD. Positive illusions and well-being revisited: Separating fact from fiction. *Psychological Bulletin* 1994;116:21–27.
21. Kesavayuth D, Poyago-Theotoky J, Tran DB, ZIkos V. Locus of control, health and healthcare utilization. *Economic Modeling* 2020;86:227–238; Botha F, Dahmann SC. Locus of control, self-control, and health outcomes. *SSM—Population Health* 2024;25:101566.

22. Moore MT, Fresco DM. Depressive realism: A meta-analytic review. *Clinical Psychology Review* 2012;32:496–509.
23. Miller K. The money-empathy gap. *New York Magazine*. June 29, 2012.
24. Piff PK, Stancato DM, Côté S, Keltner D. Higher social class predicts increased unethical behavior. *PNAS* 2012;109:4086–4091.
25. Bierce A. *The Devil's Dictionary*. Dell, 1991.
26. Loftus EF, Palmer JC. Reconstruction of automobile destruction: An example of the interaction between language and memory. *Journal of Verbal Learning and Verbal Behavior* 1974;13:585–589; Loftus EF. Leading questions and the eyewitness report. *Cognitive Psychology* 1975;7:560–572; Loftus EF, Miller DG, Burns HJ. Semantic integration of verbal information into a visual memory. *Journal of Experimental Psychology: Human Learning and Memory* 1978;4:19–31.
27. Loftus EF. Planting misinformation in the human mind: A 30-year investigation of the malleability of memory. *Learning & Memory* 2005;12:361–366.
28. Loftus EF, Davis D. Recovered memories. *Annual Review of Clinical Psychology* 2006;469–498.
29. Loftus EF. Eavesdropping on memory. *Annual Review of Clinical Psychology* 2017;68:1–18; Loftus EF. How reliable is your memory? TEDGlobal 2013; June 2013. https://www.ted.com/talks/elizabeth_loftus_how_reliable_is_your_memory?language=en
30. Simons DJ, Chabris CF. What people believe about how memory works: A representative survey of the US population. *PLoS ONE 2011*;6:e22757; Lacy JW, Stark CEL. The neuroscience of memory: Implications for the courtroom. *Nature Reviews Neuroscience* 2013;14:649–658.
31. Patihis L, Ho LY, Tingen IW, Lilienfeld SO, Loftus EF. Are the "memory wars" over? A scientist-practitioner gap in beliefs about repressed memory. *Psychological Science* 2014;25:519–530.
32. Northwestern University. Your memory is no video camera: It edits the past with present experiences. ScienceDaily. February 14, 2014. https://www.sciencedaily.com/releases/2014/02/140204185651.htm; Bridge DJ, Voss JL. Hippocampal binding of novel information with dominant memory traces can support both memory stability and change. *Journal of Neuroscience* 2014;34:2203–2213.
33. Frenda SJ, Nichols RM, Loftus EF. Current issues and advances in misinformation research. *Current Directions Psychological Science* 2011;20:20–23.
34. Kruger J, Dunning D. Unskilled and unaware of it: How difficulties in recognizing one's own competence lead to inflated self-assessments. *Journal of Personality and Social Psychology* 1999;77:1121–1134.
35. Krueger J, Mueller RA. Unskilled, unaware, or both? The contribution of social-perceptual skills and statistical regression to self-enhancement biases. *Journal of Personality and Social Psychology* 2002;27:313–327; Nuhfer E, Cogan C, Fleisher S, Gaze E, Wirth K. Random number simulations reveal how random noise affects the measurements and graphical portrayals of self-assessed competency. *Numeracy* 2016;9:article 4; Nuhfer E, Fleisher S, Cogan C, Wirth K, Gaze E. How random noise and a graphical convention subverted behavioral scientists' explanations of self-assessment data: Numeracy underlies better alternatives. *Numeracy* 2017;10:article 4.
36. Ehrlinger J, Johnson K, Banner M, Dunning D, Kruger J. Why the unskilled are unaware: Further explorations of (absent) self-insight among the incompetent. *Organizational Behavior and Human Decision Processes* 2008;105:98–121; Sanchez C, Dunning D. Overconfidence among beginners: Is a little learning a dangerous thing? *Journal of Personality and Social Psychology* 2018;114:120–128.
37. Dunning D. We are all confident idiots. *Pacific Standard*. October 27, 2014 updated June 14, 2017. https://psmag.com/social-justice/confident-idiots-92793

38 Resnick B. An expert on human blind spots gives advice on how to think. *Vox.* June 26, 2019. https://www.vox.com/science-and-health/2019/1/31/18200497/dunning-kruger-effect-explained-trump
39 Dunlosky J, Rawson KA. Overconfidence produces underachievement: Inaccurate self-evaluations undermine students' learning and retention. *Learning and Instruction* 2012;22:271–280.
40 Sanchez C, Dunning D. Overconfidence among beginners: Is a little learning a dangerous thing? *Journal of Personality and Social Psychology* 2018;114:120–128.
41 Fisher M, Keil FC. The curse of expertise: When more knowledge leads to miscalibrated explanatory insight. *Cognitive Science* 2016;40:1251–1269.
42 Son LK, Kornell N. The virtues of ignorance. *Behavioural Processes* 2020;83:207–212.

Chapter 3

1 Fisher, Goddu MK, Keil FC. Searching for explanations: How the internet inflates estimates of internal knowledge. *Journal of Experimental Psychology: General* 2015;144:674–687.
2 Collins K. The Google delusion: We're not as clever as we think we are. *Wired.* January 4, 2015. https://www.wired.com/story/google-delusion/
3 Kost A, Chen FM. Socrates was not a pimp: Changing the paradigm of questioning in medical education. *Academic Medicine* 2015;90:20–24; Priest K, King C, Chen D. Why pimping, the practice, and the word—should be eradicated from medicine. *BMC Series Blog*; October 11, 2019. http://blogs.biomedcentral.com/bmcseriesblog/2019/10/11/why-pimping-the-practice-and-the-word-should-be-eradicated-from-medicine/
4 Fisher M, Smiley AH, Grillo TLH. Information without knowledge: The effects of internet search on learning. *Memory* 2022;30:375–387.
5 Lord CG, Ross L, Lepper MR. Biased assimilation and attitude polarization: The effects of prior theories on subsequently considered evidence. *Journal of Personality and Social Psychology* 1979;37:2098–2109; Anderson CA, Lepper MR, Ross L. Perseverance of social theories: The role of explanation in the persistence of discredited information. *Journal of Personality and Social Psychology* 1980;39:1037–1049.
6 Nickerson RS. Confirmation bias: A ubiquitous phenomenon in many guises. *Review of General Psychology* 1998;2:175–220; Mahoney MJ. Publication prejudices: An experimental study of confirmatory bias in the peer review system. *Cognitive Therapy and Research* 1977;1:161–175; Hergovich A, Schott R, Burger C. Biased evaluation of abstracts depending on topic and conclusion: Further evidence of a confirmation bias within scientific psychology. *Current Psychology* 2020;29:188–209.
7 Pronin E, Lin DY, Ross L. The bias blind spot: Perceptions in self versus others. *Personality and Social Psychology Bulletin* 2002;28:369–381; Scopelliti I, Morewedge CK, McCormick E, Lauren Min H, Lebrecht S, Kassam KS. Bias blind spot: Structure, measurement, and consequences. *Management Science* 2015;61:2468–2486.
8 Pearson ML, Selby JV, Katz KA, et al. Clinical, epidemiologic, histopathologic and molecular features of an unexplained dermopathy. *PLoS ONE* 2012;7:e29908.
9 Lustig A, Mackay S, Strauss J. Morgellons disease as internet meme. *Psychosomatics* 2009;50:90; Vila-Rodriguez F, Macewan BG. Delusional parasitosis facilitated by web-based dissemination. *American Journal of Psychiatry* 2008;165:1612.
10 Middleveen MJ, Fesler MC, Stricker RB. History of Morgellons disease: From delusion to definition. *Clinical, Cosmetic and Investigational Dermatology* 2018;11:71–90.
11 Pierre JM. Gang stalking: Real-life harassment or textbook paranoia. *Psych Unseen.* October 20, 2020. https://www.psychologytoday.com/us/blog/psych-unseen/202010/gang-stalking-real-life-harassment-or-textbook-paranoia; Pierre JM. Gang stalking: Conspiracy,

delusion, and shared belief. *Psych Unseen*. October 31, 2020. https://www.psychologytoday.com/us/blog/psych-unseen/202010/gang-stalking-conspiracy-delusion-and-shared-belief; Pierre JM. Gang stalking: A case of mass hysteria. *Psych Unseen*. October 31, 2020. https://www.psychologytoday.com/us/blog/psych-unseen/202011/gang-stalking-case-mass-hysteria

12 Bell V, Maiden C, Munoz-Solomondo A, Reddy V. "Mind control" experiences on the internet: Implications for the psychiatric diagnosis of delusions. *Psychopathology* 2006;39:87–91.

13 Sheridan LP, James DV. Complaints of group-stalking ("gang-stalking"): An exploratory study of their nature and impact on complainants. *Journal of Forensic Psychiatry & Psychology* 2015;26:601–623; Sheridan L, James DV, Roth J. The phenomenology of group stalking ("gang-stalking"): A content analysis of subjective experiences. *International Journal of Environmental Research and Public Health* 2020;17:2506.

14 Lustig A, Brookes G, Hunt D. Linguistic analysis of online communication about a novel persecutory belief system (gangstalking): Mixed methods study. *Journal of Medical Internet Research* 2021;23:e25722.

15 Pierre JM. Integrating non-psychiatric models of delusion-like beliefs into forensic psychiatric assessment. *Journal of the American Academy of Psychiatry and the Law* 2019;47:171–179; Pierre JM. Forensic psychiatry versus the varieties of delusion-like belief. *Journal of the American Academy of Psychiatry and the Law* 2020;48:327–334.

16 Olson B. *Modern Esoteric: Beyond Our Senses*. Consortium of Collective Consciousness Publishing, 2018; Loftus JW. *Why I Became an Atheist: A Former Preacher Rejects Christianity*. Prometheus Books, 2012.

17 Pariser E. *The Filter Bubble: How the New Personalized Web Is Changing What We Read and How We Think*. Penguin Press, 2011.

18 New York Times. Episode two: Looking down. *Rabbit Hole*. April 23, 2020. https://www.nytimes.com/2020/04/23/podcasts/rabbit-hole-internet-youtube-virus.html

19 Straub K. www.chainsawsuit.com.

20 Brown D. *The Lost Symbol*. Doubleday, 2009.

21 McGrummen S. "Finally. Someone who thinks like me." *Washington Post*. October 1, 2016.

22 Warco K. Centerville police file additional charges against borough woman. *Observer-Reporter*. October 2018. https://www.observer-reporter.com/news/2018/oct/25/centerville-police-file-additional-charges-against-borough-woman/

23 Bakshy E, Messing S, Adamic LA. Exposure to ideologically diverse news and opinion on Facebook. *Science* 2015;348:1130–1132; Del Vicario M, Bessi A, Zollo F, et al. The spreading of information online. *Proceedings of the National Academy of Science* 2016;113:554–559; Bessi A, Zollo F, Del Vicario M, et al. Users polarization on Facebook and Youtube. *PLoS ONE* 2016;11(8):e0159641; Bail CA, Argyle LP, Brown TW, et al. Exposure to opposing views on social media can increase political polarization. *PNAS* 2018;115:9216–9221; Brugnoli E, Cinelli M, Quattrociocchi W, Scala A. Recursive patterns in online echo chambers. *Scientific Reports* 2019;9:20118; Cinelli M, Morales GDF, Galeazzi A, Quattrociocchi W, Starnini M. The echo chamber effect on social media *PNAS* 2021:118:e2023301118.

24 Flaxman S, Goel S, Rao JM. Filter bubbles, echo chambers, and online news consumption. *Public Opinion Quarterly* 2016;80:298–310; Dubois E, Blank G. The echo chamber is overstated: The moderating effect of political interest and diverse media. *Information, Communication, and Society* 2018;21:729–745.

25 Nguyen CT. Escape the echo chamber. *Aeon*. April 9, 2018. https://aeon.co/essays/why-its-as-hard-to-escape-an-echo-chamber-as-it-is-to-flee-a-cult; Nguyen CT. Echo chambers and epistemic bubbles. *Episteme* 2020;17:141–161.

26 Shugars S, Beauchamp N. Why keep arguing? Predicting engagement in political conversations online. *SAGE Open* 2019; January–March:1–13.

27 Hasell A, Weeks BE. Partisan provocation: The role of partisan news use and emotional responses in political information sharing in social media. *Human Communication Research* 2016;42:641–661; Wollebaek D, Karlsen R, Steen-Johnson K, et al. Anger, fear, and echo chambers: The emotional basis for online behavior. *Social Media + Society* 2019;5:1–14.
28 Brady WJ, McLoughlin K, Doan TN, Crockett MJ. How social learning amplifies moral outrage expression in online social networks. *Science Advances* 2021;7:eabe5641.
29 Kim JW, Guess A, Nyhan B, Reifler J. The distorting prism of social media: How self-selection and exposure to incivility fuel online comment toxicity. *Journal of Communication* 2021;71:922–946.
30 Rathje S, Van Bavel JJ, van der Linden S. Out-group animosity drives engagement on social media. *PNAS* 2021;118:e2024292118.
31 Munn L. Angry by design: Toxic communication and technical architectures. *Humanities and Social Sciences Communications* 2020;7:53.
32 Milli S, Carroll M, Pandey S, Wang Y, Dragan AD. Twitter's algorithm: Amplifying anger, animosity, and affective polarization. *arXiv* 2305.16941.
33 Suler J. The online disinhibition effect. *Cyberpsychology and Behavior* 2004;7:321–326.
34 Haines R, Hough J, Cao L, et al. Anonymity in computer-mediated communication: More contrarian ideas with less influence. *Group Decision and Negotiation* 2014;23:765–786.
35 Santana AD. Virtuous or vitriolic: The effect of anonymity on civility in online newspaper reader comment boards. *Journalism Practice* 2014;8:18–33.
36 Anderson AA, Brossard D, Scheufele DA, et al. The "nasty effect": Online incivility and risk perceptions of emerging technologies. *Journal of Computer-Mediated Communication* 2014;19:373–387.
37 Vaidhyanathan S. *Anti-Social Media*. Oxford University Press, 2018.
38 Rathje S, Robertson C, Brady WJ, Van Bavel JJ. People think that social media platforms do (but should not) amplify divisive content. *Perspectives on Psychological Science* 2023. https://doi.org/10.1177/17456916231190

Chapter 4

1 Rodia T. Is it a cult, or a new religious movement? *Penn Today*. August 29, 2019. https://penntoday.upenn.edu/news/it-cult-or-new-religious-movement; Pierre JM. Cults of personality. In: Scott C, McDermott B, eds. *A Clinical Guide to Cults and Persuasive Leadership*. Cambridge University Press, in press.
2 Bearman J. Heaven's Gate: The sequel. *LA Weekly*. March 21, 2007. https://www.laweekly.com/heavens-gate-the-sequel/
3 Niebuhr G. On the furthest fringes of millennialism. *The New York Times*. March 28, 1997. https://archive.nytimes.com/www.nytimes.com/library/national/mass-suicide-cult.html; Abromyaityte M. The 1990s cult "Heaven's Gate" has four remaining followers—we spoke to them. *Vice*. October 15, 2020. https://www.vice.com/en/article/v7gjky/heavens-gate-cult-remaining-members
4 Hafford M. Heaven's Gate 20 years later: 10 things you didn't know. *Rolling Stone*. March 24, 2017. https://www.rollingstone.com/feature/heavens-gate-20-years-later-10-things-you-didnt-know-114563/
5 Associated Press. Mass suicide "a good way to get rid of a few nuts," Turner says. *The Spokesman-Review*. March 30, 1997. https://www.spokesman.com/stories/1997/mar/30/mass-suicide-a-good-way-to-get-rid-of-a-few-nuts/
6 European Southern Observatory. Fraudulent use of a Ifa/UH picture. https://www.eso.org/~ohainaut/Hale_Bopp/hb_ufo_tholen.html

7. Genoni T. Art Bell, Heaven's Gate, and journalistic integrity. *Skeptical Inquirer* 1997;21:22–23.
8. Aslett K, Sanderson Z, Godel W, Persily N, Nagler J, Tucker JA. Online searches to evaluate misinformation can increase its perceived veracity. *Nature* 2023. https://doi.org/10.1038/s41586-023-06883-y
9. Goldman A. The Comet Ping Pong gunman answers our reporter's questions. *The New York Times*. December 7, 2016. https://www.nytimes.com/2016/12/07/us/edgar-welch-comet-pizza-fake-news.html
10. Spiegelman I. Daredevil and flat-earther "Mad Mike" Hughes dies in homemade rocket crash. *Los Angeles Magazine*. February 24, 2020. https://lamag.com/science/mad-mike-hughes-rocket-crash; Ortiz A. Mike Hughes, 64, D.I.Y. daredevil, is killed in rocket crash. *The New York Times*. February 23, 2020. https://www.nytimes.com/2020/02/23/us/mad-mike-hughes-dead.html
11. Cohen J. Covid-19 vaccine hesitancy is worse in the E.U. than U.S. *Forbes*. March 8, 2021. https://www.forbes.com/sites/joshuacohen/2021/03/08/covid-19-vaccine-hesitancy-is-worse-in-eu-than-us/?sh=386f0529611f
12. Johnson CK, Stobbe M. Nearly all COVID deaths in the US are now among unvaccinated. *AP News*. June 29, 2021. https://apnews.com/article/coronavirus-pandemic-health-941fcf43d9731c76c16e7354f5d5e187
13. Scobie HM, Johnson AG, Suthar AB, et al. Monitoring incidence of COVID-19 cases, hospitalizations, and deaths, by vaccination status—13 U.S. Jurisdictions, April 4–July 17, 2021. *Morbidity Mortal Weekly Report* 2021;70:1284–1290; Johnson AG, Linde L, Ali AR, et al. COVID-19 incidence and mortality among unvaccinated and vaccinated persons aged ≥ 12 years by receipt of bivalent booster doses and time since vaccination—24 U.S. jurisdictions, October 3, 2021–December 24, 2022. *Morbidity Mortal Weekly Report* 2023;72:145–152.
14. Simmons-Duffin S, Nakajima K. This is how many lives could have been saved with COVID vaccinations in each state. NPR.org. May 13, 2022. https://www.npr.org/sections/health-shots/2022/05/13/1098071284/this-is-how-many-lives-could-have-been-saved-with-covid-vaccinations-in-each-state; and Vaccine preventable death analysis. Globalepidemics.org. Data from January 2021–April 2022. https://globalepidemics.org/vaccinations/
15. Mitchell A, Jurkowitz M, Oliphant JB, Shearer E. Americans who mainly get their news on social media are less engaged, less knowledgeable. Pew Research Center. July 30, 2020. https://www.pewresearch.org/journalism/2020/07/30/americans-who-mainly-get-their-news-on-social-media-are-less-engaged-less-knowledgeable/
16. Matsa KA. Fewer Americans rely on TV news; what type they watch varies by who they are. Pew Research Center. January 5, 2018. https://www.pewresearch.org/fact-tank/2018/01/05/fewer-americans-rely-on-tv-news-what-type-they-watch-varies-by-who-they-are/; Pew Research Center. Network news fact sheet. Journalism.org July 13, 2021. https://www.journalism.org/fact-sheet/network-news/; Pew Research Center. Cable news fact sheet. July 13, 2021. https://www.journalism.org/fact-sheet/cable-news/.
17. Hruby P. The SportsCenter-ization of political journalism. *The Atlantic*. January 4, 2012. https://www.theatlantic.com/politics/archive/2012/01/the-sportscenter-ization-of-political-journalism/250882/
18. The O'Reilly Factor. March 2, 2016. https://www.youtube.com/watch?v=slC2DaFMRTw
19. Jones JM. U.S. media trust continues to recover from 2016 low. Gallup. October 12, 2018. https://news.gallup.com/poll/243665/media-trust-continues-recover-2016-low.aspx
20. Koppel T. The case against news we can choose. *The Washington Post*. November 14, 2010. https://www.washingtonpost.com/archive/opinions/2010/11/14/the-case-against-news-we-can-choose/8818803c-f580-11df-a418-19c210d82b3f/

21 The Late Show with Stephen Colbert. November 24, 2015. https://www.youtube.com/watch?v=z6KnZTRw6ZU
22 Council on Foreign Relations. A conversation with Ted Koppel. November 12, 2019. Transcript available at: https://www.cfr.org/event/distinguished-voices-series-ted-koppel
23 Moyer JW. Trump inspires gloomy Ted Koppel to scold Bill O'Reilly over the state of TV news. *The Washington Post.* March 3, 2016. https://www.washingtonpost.com/news/morning-mix/wp/2016/03/03/trump-inspires-gloomy-ted-koppel-to-scold-bill-oreilly-over-the-state-of-tv-news/
24 Polman D. The "objective" journalistic nirvana. WHYY.org/News. November 17, 2010. https://whyy.org/articles/the-qobjectiveq-journalistic-nirvana/
25 Madison J. The Federalist Papers: No. 10. November 23, 1787. https://avalon.law.yale.edu/18th_century/fed10.asp
26 Dobski B. America is a republic, not a democracy. *First Principles* 80; June 2020; Thomas G. "America is a republic, not a democracy" is a dangerous—and wrong—argument. *The Atlantic.* November 2, 2020. https://www.theatlantic.com/ideas/archive/2020/11/yes-constitution-democracy/616949/
27 Lemann N. Amateur hour. *The New Yorker.* August 7, 2006.
28 Kavanagh J, Rich MD. *Truth Decay: An Initial Exploration of the Diminishing Role of Facts and Analysis in American Public Life.* RAND Corporation, 2018.
29 Ruane KA. Fairness doctrine: History and Constitutional issues. Congressional Research Service Report for Congress. July 13, 2011; Matthews D. Everything you need to know about the Fairness Doctrine in one post. *The Washington Post.* August 23, 2011.
30 Berg M, Brown A. The highest paid YouTube stars of 2020. *Forbes.* December 18, 2020. https://www.forbes.com/sites/maddieberg/2020/12/18/the-highest-paid-youtube-stars-of-2020/
31 Vosoughi S, Roy D, Aral S. The spread of true and false news online. *Science* 2018;359:1146–1151.
32 Aral S. How lies spread online. *The New York Times.* March 8, 2018. https://www.nytimes.com/2018/03/08/opinion/sunday/truth-lies-spread-online.html
33 Taylor LE, Swerdfeger AL, Eslick CD. Vaccines are not associated with autism: An evidence-based meta-analysis of case-control and cohort studies. *Vaccine* 2014;32:3623–3629.
34 Kata A. A postmodern Pandora's box: Anti-vaccination misinformation on the internet. *Vaccine* 2010;28:1709–1716.
35 Elkin LE, Pullon SRH, Stubbe MH. "Should I vaccinate my child?" comparing the displayed stances of vaccine information retrieved from *Google, Facebook,* and *YouTube. Vaccine* 2020;38:2771–2778.
36 Johnson NF, Velásquez N, Restrepo NJ, et al. The online competition between pro- and anti-vaccination views. *Nature* 2020;582:230–233.
37 Frankovic K. Why won't Americans get vaccinated? YouGov.com. July 15, 2021. https://today.yougov.com/topics/politics/articles-reports/2021/07/15/why-wont-americans-get-vaccinated-poll-data
38 Lazer D, Green J, Ognyanova K, et al. The COVID States Project #57: Social media news consumption and COVID-19 vaccination rates. July 2021. https://osf.io/uvqbs/
39 Cathey L. President Biden says Facebook, other social media "killing people" when it comes to COVID-19 misinformation. ABC News. July 16, 2021. https://abcnews.go.com/Politics/president-biden-facebook-social-media-killing-people-covid/story?id=78890692
40 Wang Y, McKee M, Torbica A, Stuckler D. Systematic literature review on the spread of health misinformation on social media. *Social Science & Medicine* 2019;240:112552.
41 Johnson SB, Parsons M, Dorff T, et al. Cancer misinformation and harmful information on Facebook and other social media: A brief report. *Journal of the National Cancer Institute* 2022; 114:1036–1039.

42 Beall J. Medical publishing and the threat of predatory journals. *International Journal of Women's Dermatology* 2016;2:115–116.
43 Eriksson S, Helgesson G. Time to stop talking about "predatory journals." *Learned Publishing* 2018;31:181–183.
44 Beall J. Predatory publishers are corrupting open access. *Nature* 2012;489:179.
45 Beall J. The open-access movement is not really about open access. *tripleC* 2013;11:589–597; Gillis AG. The rise of junk science. *The Walrus*. May 27, 2019, updated December 6, 2021. https://thewalrus.ca/the-rise-of-junk-science/
46 Lawton G. Science in crisis. *New Scientist*. 2020;246:12–14.
47 Stefansky E. Watch Ted Koppel tell Sean Hannity that he's "bad for America." *Vanity Fair*. March 26, 2017.
48 City News Service. San Diego judge dismisses OAN's $10 million defamation lawsuit against Rachel Maddow. KPBS.org. May 23, 2020. https://www.kpbs.org/news/2020/05/23/san-diego-judge-dismisses-oans-10-million-defamati
49 Folkenflik D. You literally can't believe the facts Tucker Carlson tells you. So say Fox's lawyers. NPR.org. September 29, 2020. https://www.npr.org/2020/09/29/917747123/you-literally-cant-believe-the-facts-tucker-carlson-tells-you-so-say-fox-s-lawye
50 Stanford History Education Group. Evaluating information: The cornerstone of civic online reasoning. 2016. https://stacks.stanford.edu/file/druid:fv751yt5934/SHEG%20Evaluating%20Information%20Online.pdf
51 Morris A, Brading H. E-literacy and the grey digital divide: A review with recommendations. *Journal of Information Literacy* 2007;1:13–28; Guess A, Nagler J, Tucker J. Less than you think: Prevalence and predictors of fake new dissemination on Facebook. *Science Advances*; 2019;5:eaau4586; Grinberg N, Joseph K, Friedland L, Swire-Thompson B, Lazer D. Fake news on Twitter during the 2016 U.S. presidential election. *Science* 2019;363:374–378.
52 Alfano S. The truth of truthiness. *CBSNews.com*. December 12, 2006. https://www.cbsnews.com/news/the-truth-of-truthiness/; Stephen Colbert, *The Colbert Report*, Comedy Central. October 17, 2005. https://www.cc.com/video/63ite2/the-colbert-report-the-word-truthiness; American Dialect Society. Truthiness voted 2005 word of the year, AmericanDialect.org. January 6, 2006. https://americandialect.org/truthiness_voted_2005_word_of_the_year.
53 Kahan DM. The politically motivated reasoning paradigm part 1: What is politically motivated reasoning and how to measure it. In: Scott R, Kosslyn S, eds. *Emerging Trends in the Social and Behavioral Sciences*. Wiley, 2016; Kahan DM. The politically motivated reasoning paradigm part 2: Unanswered questions. In: Scott R, Kosslyn S, eds. *Emerging Trends in the Social and Behavioral Sciences*. Wiley, 2016.
54 Taber CS, Lodge CS. Motivated skepticism in the evaluation of political beliefs. *American Journal of Political Science* 2006;50:755–769; Campbell TH, Kay AC. Solution aversion: On the relation between ideology and motivated disbelief. *Journal of Personality and Social Psychology* 2014;107(5):809–824; Williams D. Motivated ignorance, rationality, and democratic politics. *Synthese* 2020.
55 Kahan DM. Misconceptions, misinformation, and the logic of identity-protective cognition. Cultural Cognition Project Working Paper Series No. 164. May 24, 2017; Van Bavel JJ, Pereira A. The partisan brain: An identity-based model of political belief. *Trends in Cognitive Science* 2018;22:213–224.
56 Goldberg M. What Fox News says when you're not listening. *The New York Times*. February 17, 2023; Darcy O. Fox News stars and executives privately trashed Trump's election fraud claims, court document reveals. CNN.com. February 17, 2023. https://www.cnn.com/2023/02/16/media/fox-news-stars-executives-court-documents/index.html
57 Kunda Z. The case for motivated reasoning. *Psychological Bulletin* 1990;480–498.

58 Kahan DM, Jenkins-Smith H, Braman D. Cultural cognition of scientific consensus. *Journal of Risk Research* 2011;14:147–174; Michael RB, Breaux BO. The relationship between political affiliation and beliefs about sources of "fake news." *Cognitive Research: Principles and Implications* 2021;6:6.
59 Asimov I. A cult of ignorance. *Newsweek.* January 21, 1980; Sagan C. *The Demon-Haunted World: Science as a Candle in the Dark.* Ballantine Books, 1995.
60 Stanovich KE, West RF, Toplak ME. Myside bias, rational thinking, and intelligence. *Current Directions in Psychological Science* 2013;22:259–264.
61 Kahan DM, Peters E, Dawson EC, Slovic P. Motivated numeracy and enlightened self-government. *Behavioural Public Policy* 2017;1:54–86; Kahan DM. Ideology, motivated reasoning, and cognitive reflection. *Judgment and Decision Making* 2013;8:407–424.
62 Pennycook G, Rand DG. Lazy, not biased: Susceptibility to partisan fake news is better explained by lack of reasoning than by motivated reasoning. *Cognition* 2019;188:39–50; Bago B, Rand DG, Pennycook G. Fake news, fast and slow: Deliberation reduces belief in false (but not true) news headlines. *Journal of Experimental Psychology: General* 2020;149:1608–1613.
63 Tappin BM, Pennycook G, Rand DG. Thinking clearly about causal inferences of politically motivated reasoning: Why paradigmatic study designs often undermine causal inference. *Current Opinion in Behavioral Science* 2020:34:81–87; Tappin BM, Pennycook G, Rand DG. Rethinking the link between cognitive sophistication and politically motivated reasoning. *Journal of Experimental Psychology: General* 2021;150:1095–1114.
64 Bullock JG, Gerber AS, Hills SJ, Huber GA. Partisan bias in factual beliefs about politics. *Quarterly Journal of Political Science* 2015;10:519–578.

Chapter 5

1 Antidefamation League. "The lawless ones": The resurgence of the sovereign citizen movement, 2nd ed. Antidefamation League Special Report. 2012. https://www.adl.org/sites/default/files/documents/assets/pdf/combating-hate/Lawless-Ones-2012-Edition-WEB-final.pdf; Berger JM. Without prejudice: What sovereign citizens believe. Program on Extremism at George Washington University. June 2016. https://extremism.gwu.edu/sites/g/files/zaxdzs2191/f/downloads/JMB%20Sovereign%20Citizens.pdf; Southern Poverty Law Center. Sovereign Citizen Movement. https://www.splcenter.org/fighting-hate/extremist-files/ideology/sovereign-citizens-movement
2 Pierre JM. Forensic psychiatry versus the variety of delusion-like beliefs. *Journal of the American Academy of Psychiatry and the Law* 2020;48:327–334.
3 Seifert C. The distributed influence of misinformation. *Journal of Applied Research in Memory and Cognition* 2017;397–400.
4 Hardwig J. The role of trust in knowledge. *Journal of Philosophy* 1991;88:693–708; Sperber D, Clément F, Heintz C, Mascaro O, Mercier H, Origgi G, Wilson D. Epistemic vigilance. *Mind & Language* 2010;25:359–393.
5 Kareklas I, Muehling DD, Weber TJ. Reexamining health messages in the digital age: A fresh look at source credibility effects. *Journal of Advertising* 2015;44:88–104.
6 Jones JM. U.S. media trust continues to recover from 2016 low. Gallup. October 12, 2018. https://news.gallup.com/poll/243665/media-trust-continues-recover-2016-low.aspx
7 Pew Research Center. Public trust in government: 1958–2021. May 17, 2021. https://www.pewresearch.org/politics/2021/05/17/public-trust-in-government-1958-2021/
8 Kahan DM. Misconceptions, misinformation, and the logic of identity-protective cognition. Cultural Cognition Project Working Paper Series No. 164. May 24, 2017.

9. Sunstein CR, Vermeule A. Conspiracy theories: Causes and cures. *Journal of Political Philosophy* 2009;17:202–227.
10. Rosselli R, Martini M, Bragazzi NL. The old and the new: Vaccine hesitancy in the era of the Web 2.0. Challenges and opportunities. *Journal of Preventive Medicine and Hygiene* 2006;57:E47–E50.
11. Stelter B. Newsmax TV scores a ratings win over Fox News for the first time ever. *CNN.com*. December 8, 2020. https://www.cnn.com/2020/12/08/media/newsmax-fox-news-ratings/index.html
12. Williamson E, Steel E. Conspiracy theories made Alex Jones very rich. They may bring him down. *The New York Times*. September 7, 2018. https://www.nytimes.com/2018/09/07/us/politics/alex-jones-business-infowars-conspiracy.html
13. Murdock S. Alex Jones' Infowars store made $165 million over 3 years, records show. *Huffington Post*. January 7, 2022. https://www.huffpost.com/entry/infowars-store-alex-jones_n_61d71d8fe4b0bcd2195c6562
14. Siemaszko C. InfoWars' Alex Jones is a "performance artist," his lawyer says in a divorce hearing. *NBCNews.com*. April 17, 2017. https://www.nbcnews.com/news/us-news/not-fake-news-infowars-alex-jones-performance-artist-n747491
15. Maxouris C, Joseph E. Alex Jones says "form of psychosis" made him believe events like Sandy Hook massacre were staged. *CNN.com*. April 1, 2019. https://www.cnn.com/2019/03/30/us/alex-jones-psychosis-sandy-hook/index.html
16. Jones A. 45 battles UN offensive plus Alex Jones' lawyer tells all. *The Alex Jones Show*. March 31, 2019. https://banned.video/watch?id=5ca157e04b025c001747b14b
17. Zaitchik A. Meet Alex Jones. *Rolling Stone*. March 2, 2011. https://www.rollingstone.com/culture/culture-news/meet-alex-jones-175845/
18. Brown S. Alex Jones' media empire is a machine built to sell snake-oil diet supplements. *New York Magazine*. May 4, 2017. https://nymag.com/intelligencer/2017/05/how-does-alex-jones-make-money.html
19. Levy SG. Everything you need to know about Goop's jade-egg lawsuit. *Vogue.com*. September 5, 2018. https://www.vogue.com/article/goop-jade-yoni-egg-lawsuit-gwyneth-paltrow-vaginal-pelvic-floor-health
20. Dickson EJ. We fact-checked four of the most outrageous claims in Gwyneth Paltrow's Netflix show. *Rolling Stone,* January 29, 2020. https://www.rollingstone.com/culture/culture-features/gwyneth-paltrow-goop-lab-netflix-941830/; St. Felix D. The magical thinking of "The Goop Lab." *The New Yorker*. February 3, 2003. https://www.newyorker.com/magazine/2020/02/03/the-magical-thinking-of-the-goop-lab
21. Khazan O. The baffling rise of Goop. *The Atlantic*. September 12, 2017. https://www.theatlantic.com/health/archive/2017/09/goop-popularity/539064/
22. Merrill P. Why the haters mean nothing to Gwyneth Paltrow. *CEO Magazine*. May 26, 2021. https://www.theceomagazine.com/business/health-wellbeing/gwyneth-paltrow-goop/
23. Funk C, Kennedy B. Public confidence in scientists has remained stable for decades. Pew Research Center. August 27, 2020. https://www.pewresearch.org/fact-tank/2020/08/27/public-confidence-in-scientists-has-remained-stable-for-decades/
24. Team Goop. Uncensored: A word from our contributing doctors. https://goop.com/wellness/health/uncensored-a-word-from-our-doctors/
25. Mull A. I Gooped myself. *The Atlantic*. August 26, 2019.
26. Stamp N. Gwyneth Paltrow's "Goop Lab" is horrible. The medical industry is partly to blame. *The Washington Post*. February 8, 2020. https://www.washingtonpost.com/opinions/2020/02/08/gwyneth-paltrows-goop-lab-is-horrible-medical-industry-is-partly-blame/
27. Johnson SB, Park HS, Gross CP, Yu JB. Complementary medicine, refusal of conventional cancer therapy, and survival among patients with curable cancers. *JAMA Oncology* 2018;4:1375–1381.

28 Remski M. Inside Kelly Brogan's COVID-denying, vax-resistant conspiracy machine. *Medium*. September 15, 2020. https://gen.medium.com/inside-kelly-brogans-covid-denying-vax-resistant-conspiracy-machine-28342e6369b1
29 Abbott RD, Sherwin K, Klopf H, Mattingly HJ, Brogan K. Efficacy of a multimodal online lifestyle intervention for depressive symptoms and quality of life in individuals with a history of major depressive disorder. *Cureus* 2020;12(7):e9061.
30 Frankel J. HIV doesn't cause AIDS according to Gwyneth Paltrow Goop "trusted expert" doctor Kelly Brogan. *Newsweek*. December 6, 2017. https://www.newsweek.com/hiv-doesnt-cause-aids-according-gwyneth-paltrow-goop-doctor-kelly-brogan-735645
31 Spiegelman I. A controversial psychiatrist and Goop contributor suggests that coronavirus isn't real. *Los Angeles Magazine*. March 24, 2020. https://lamag.com/health/kelly-brogan-coronavirus
32 The Chalkboard Editorial Team. Dr. Kelly Brogan's jaw-dropping interview on holistic psychiatry. Thechalkboardmag.com. November 12, 2019. https://thechalkboardmag.com/kelly-brogan-holistic-psychiatry
33 Center for Countering Digital Hate. Why platforms must act on twelve leading online anti-vaxxers. March 24, 2021. https://cdn.centerforinquiry.org/wp-content/uploads/sites/33/2021/05/15131138/disinformation_dozen.pdf
34 Smith MR, Reiss J. Inside one network cashing in on vaccine disinformation. *AP News*. May 13, 2021. https://apnews.com/article/anti-vaccine-bollinger-coronavirus-disinformation-a7b8e1f33990670563b4c469b462c9bf
35 US Food and Drug Administration. Warning letter to Dr. Joseph Mercola, Dr. Mercola's Natural Health Center. March 22, 2011. https://web.archive.org/web/20111208202236/http:/www.fda.gov/ICECI/EnforcementActions/WarningLetters/2011/ucm250701.htm; US Food and Drug Administration. Warning letter to Dr. Joseph Mercola, Mercola.com, LLC. February 18, 2021. https://www.fda.gov/inspections-compliance-enforcement-and-criminal-investigations/warning-letters/mercolacom-llc-607133-02182021; Evans J. $2.59 million in refunds for Mercola tanning beds. consumer.ftc.gov. February 7, 2017. https://www.consumer.ftc.gov/blog/2017/02/259-million-refunds-mercola-tanning-beds
36 Frenkel S. The most influential spreader of coronavirus misinformation online. *The New York Times*. July 24, 2021. https://www.nytimes.com/2021/07/24/technology/joseph-mercola-coronavirus-misinformation-online.html
37 Smith B. Dr. Mercola: visionary or quack? *Chicago Magazine*. January 31, 2012. https://www.chicagomag.com/chicago-magazine/february-2012/dr-joseph-mercola-visionary-or-quack/
38 Subramanian S. Inside the Macedonian fake-news complex. *Wired*. February 25, 2017. https://www.wired.com/2017/02/veles-macedonia-fake-news/
39 Hughes HC, Waismel-Manor I. The Macedonian fake news industry and the 2016 US election. *PS: Political Science and Politics* 2021;54:19–23.
40 Broniatowski DA, Jamison AM, Q, S, AlKulaib, L, et al. Weaponized health communication: Twitter bots and Russian trolls amplify the vaccine debate. *American Journal of Public Health* 2018;108;1378–1384.
41 Dunn AG, Surian D, Dalmazzo J, et al. Limited role of bots in spreading vaccine-critical information among active Twitter users in the United States: 2017–2019. *American Journal of Public Health* 2020;110(S3):S319–S325.
42 Spitale G, Biller-Andorno N, Federico G. AI model GPT-3 (dis)informs us better than humans. *Science Advances* 2023;9:eadh1850.
43 Hasher L, Goldstein D, Topping T. Frequency and the conference of referential validity. *Journal of Verbal Learning and Verbal Behavior* 1977;16:107–112.
44 Reber R, Schwarz N. Effects of perceptual fluency on judgments of truth. *Consciousness and Cognition: An International Journal* 1999;8:338–342; McGlone MS, Tofighbakhsh J.

Birds of a feather flock conjointly (?): Rhyme as reason in aphorisms. *Psychological Science* 2000;11:424–428.
45 Dechene A, Stahl C, Hansen J, Wanke M. The truth about the truth: A meta-analytic review of the truth effect. *Personality and Social Psychology Review* 2919;14:238–257.
46 Arkes HR, Hackett C, Boehm L. The generality of the relation between familiarity and judged validity. *Journal of Behavioral Decision Making* 1989;2:81–94.
47 Wegner DM, Wenzlaff R, Kerker RM, Beattie AE. Incrimination through innuendo: Can media questions become public answers? *Journal of Personality and Social Psychology* 1981;40:5:822–832.
48 Fazio LK, Brashier NM, Payne BK, Marsh EJ. Knowledge does not protect against illusory truth. *Journal of Experimental Psychology: General* 2015;144:993–1002.
49 Pennycook G, Cannon TD, Rand DG. Prior exposure increases perceived accuracy of fake news. *Journal of Experimental Psychology: General* 2018;147:1865–1880.
50 Orwell G. *1984*. Harcourt Brace Jovanovich, 1949.
51 Paul C, Matthews M. *The Russian "Firehose of Falsehood" Propaganda Model: Why It Might Work and Options to Counter It*. RAND Corporation, 2016. https://www.rand.org/pubs/perspectives/PE198.html
52 Shonam S. "Our task was to set Americans against their own government": New details emerge about Russia's trolling operation. *Business Insider*. October 17, 2017. https://www.businessinsider.com/former-troll-russia-disinformation-campaign-trump-2017-10
53 Linvill DL, Warren PL. Troll factories: Manufacturing specialized disinformation on Twitter. *Political Communication* 2020;37:447–467.
54 Grinberg N, Joseph K, Friedland L, Swire-Thompson B, Lazer D. Fake news on Twitter during the 2016 U.S. presidential election. *Science* 2019;363:374–378; Guess AM, Nyhan B, Reifler J. Exposure to untrustworthy websites in the 216 US election. *Nature Human Behavior* 2020;4:472–480; Bail CA, Guay B, Maloney E, et al. Assessing the Russian Internet Research Agency's impact on the political attitudes and behaviors of American Twitter users in late 2017. *Proceedings of the National Academy of Sciences* 2020;117:243–250.
55 Howard PN, Ganesh B, Liotsiou D, Kelly J, François C. The IRA, social media and political polarization in the United States, 2012–2018. Oxford Internet Institute, University of Oxford 2018. https://digitalcommons.unl.edu/senatedocs/1/
56 Bradshaw S, Howard PN. The global disinformation disorder: 2019 global inventory of organized social media manipulation. Working Paper 2019-2. Oxford UK. Project on Computational Propaganda. 2019.
57 Meleschevich K, Schafer B. Online information laundering: the role of social media. Policy Brief, Alliance for Securing Democracy. January 9, 2018. https://securingdemocracy.gmfus.org/online-information-laundering-the-role-of-social-media/
58 Al-Rawi, Rahman A. Manufacturing rage: The Russian Internet Research Agency's political astroturfing on social media. *First Monday* 2020;25.
59 Illing S. The Russian roots of our misinformation problem. *Vox*. October 26, 2020. https://www.vox.com/world/2019/10/24/20908223/trump-russia-fake-news-propaganda-peter-pomerantsev
60 Oliver JE, Wood TJ. Conspiracy theories and the paranoid style(s) of mass opinion. *American Journal of Political Science* 2014;58, 952–966.
61 Holmes J. A Trump surrogate drops the mic: "There's no such thing as facts." *Esquire*. December 1, 2016. https://www.esquire.com/news-politics/videos/a51152/trump-surrogate-no-such-thing-as-facts/
62 Robertson L, Farley R. The facts on crowd size. Factcheck.org. January 23, 2017. https//www.factcheck.org/2017/01/the-facts-on-crowd-size/
63 Sinderbrand R. How Kellyanne Conway ushered in the era of "alternative facts." *The Washington Post*. January 22, 2017. https://www.washingtonpost.com/news/the-fix/wp/2017/01/22/how-kellyanne-conway-ushered-in-the-era-of-alternative-facts/

64 Kessler G, Rizzo S, Kelly M. Trump's false or misleading claims total 30,573 over 4 years. *The Washington Post.* January 24, 2021. https://www.washingtonpost.com/politics/2021/01/24/trumps-false-or-misleading-claims-total-30573-over-four-years/
65 Parker A. Spin, hyperbole and deception: How Trump claimed credit for an Obama veterans achievement. *The Washington Post.* October 23, 2020. https://www.washingtonpost.com/politics/2020/10/23/trump-obama-veterans-choice-act/
66 Associated Press. Transcript of Trump's speech at rally before US Capitol riot. Apnews.com. January 13, 2021. https://apnews.com/article/election-2020-joe-biden-donald-trump-capitol-siege-media-e79eb5164613d6718e9f4502eb471f27
67 Monmouth University Polling Institute. Public supports both early voting and requiring photo ID to vote. Monmouth University Poll. June 21, 2021. https://www.monmouth.edu/polling-institute/reports/monmouthpoll_us_062121/
68 Carbonaro G. 40% of Americans think 2020 election was stolen, just days before midterms. Newsweek.com. November 2, 2022. https://www.newsweek.com/40-americans-think-2020-election-stolen-days-before-midterms-1756218; Durkee A. Republicans increasingly realize there's no evidence of election fraud—but most still think 2020 election was stolen anyway, poll finds. Forbes. March 14, 2023. https://www.forbes.com/sites/alisondurkee/2023/03/14/republicans-increasingly-realize-theres-no-evidence-of-election-fraud-but-most-still-think-2020-election-was-stolen-anyway-poll-finds
69 Nuzzi O. Kellyanne Conway is a star. *New York Magazine.* March 18, 2017. https://nymag.com/intelligencer/2017/03/kellyanne-conway-trumps-first-lady.html
70 Arendt H. Hannah Arendt: From an interview. *New York Times Review of Books.* October 26, 1978. https://www.nybooks.com/articles/1978/10/26/hannah-arendt-from-an-interview/
71 Lewandowsky S, Ecker UKH, Cook J. Beyond misinformation: Understanding and coping with the "post-truth" era. *Journal of Applied Research in Memory and Cognition* 2017;6:353–369.
72 Newall M. Misinformation around U.S. Capitol unrest, election spreading among Americans. Ipsos Poll. January 19, 2021. https://www.ipsos.com/en-us/news-polls/misinformation-unrest-election-spreading
73 Nichols T. *The Death of Expertise: The Campaign against Established Knowledge and Why It Matters.* Oxford University Press, 2017.

Chapter 6

1 Clifton A, Frye C, Jefferson R. Kyrie Irving—DEEP in thought 30,000 feet high above. Road Trippin. February 16, 2017. https://www.youtube.com/watch?v=mzjL9JxSFAk
2 Martin D. Charles Johnson, 76, proponent of flat Earth. *The New York Times.* March 25, 2001. https://www.nytimes.com/2001/03/25/us/charles-johnson-76-proponent-of-flat-earth.html
3 Nguyen H. Most flat earthers consider themselves very religious. YouGov. April 2, 2018. https://today.yougov.com/topics/society/articles-reports/2018/04/02/most-flat-earthers-consider-themselves-religious; Picheta R. The flat-Earth conspiracy is spreading around the globe. Does it hide a darker core? CNN.com. November 18, 2019. https://www.cnn.com/2019/11/16/us/flat-earth-conference-conspiracy-theories-scli-intl/index.html
4 Said-Moorhouse L. Rapper B.o.B thinks the Earth is flat, has photographs to prove it. CNN.com. January 26, 2016. https://www.cnn.com/2016/01/26/entertainment/rapper-bob-earth-flat-theory/; Ramisetti K. Tila Tequila goes on bizarre Twitter rant, insists Earth if flat and declares she's immortal. *New York Daily News.* January 7, 2016. https://www.nydailynews.com/entertainment/gossip/tila-tequila-insists-earth-flat-bizarre-twitter-rant-article-1.2489596
5 Pierre JM. Mistrust and misinformation: A two component, socio-epistemic model of belief in conspiracy theories. *Journal of Social and Political Psychology* 2020;8:617–641.

6   Neuharth-Keusch AJ. Cavaliers All-Star Kyrie Irving legitimately believes the Earth is flat. USA Today. February 18, 2017. https://www.usatoday.com/story/sports/nba/allstar/2017/02/18/kyrie-irving-earth-flat-nba-all-star/98090104/
7   ESPN. Kyrie Irving apologizes for saying Earth is flat: "Didn't realize the effect." ESPN.com. October 1, 2018. https://www.espn.com/nba/story/_/id/24863899/kyrie-irving-boston-celtics-apologizes-saying-earth-flat
8   Deb S. Kyrie Irving doesn't know if the Earth is round or flat. He does want to discuss it. *The New York Times*. June 8, 2018. https://www.nytimes.com/2018/06/08/movies/kyrie-irving-nba-celtics-earth.html
9   Dean S. No, one-third of Millennials don't actually think Earth is flat. *Science Alert*. April 4, 2018. https://www.sciencealert.com/one-third-millennials-believe-flat-earth-conspiracy-statistics-yougov-debunk
10 Goertzel T. Belief in conspiracy theories. *Political Psychology* 1994;15:731–742; Oliver JE, Wood, TJ. Conspiracy theories and the paranoid style(s) of mass opinion. *American Journal of Political Science* 2014;58:952–966; Oliver JE, Wood TJ. Medical conspiracy theories and health behaviors in the United States. *JAMA Internal Medicine* 2014;174:817–818; Mancuso M, Vassallo S, Vezzoni C. Believing in conspiracy theories: Evidence from an exploratory analysis of Italian survey data. *South European Society and Politics* 2017;3:327–344.
11 YouGov. YouGov Cambridge Globalism Project—Conspiracy theories. YouGov Poll. February 28–March 26, 2019. https://d25d2506sfb94s.cloudfront.net/cumulus_uploads/document/2c6lta5kbu/YouGov%20Cambridge%20Globalism%20Project%20-%20Conspiracy%20Theories.pdf
12 Goertzel T. Belief in conspiracy theories. *Political Psychology* 1994;15:731–742; Swami V, Coles R, Stieger S, et al. Conspiracist ideation in Britain and Austria: Evidence of a monological belief system and associations between individual psychological differences and real-world and fictitious conspiracy theories. *British Journal of Psychology* 2011;102:443–463; Wood MJ, Douglas KM, Sutton RM. Dead and alive: Beliefs in contradictory conspiracy theories. *Social Psychology and Personality Science* 2012;3:767–773.
13 Uscinski J, Parent J. *American Conspiracy Theories*. Oxford University Press, 2014.
14 Uscinski J, Enders A, Klofstad C, et al. Have beliefs in conspiracy theories increased over time? *PLoS ONE* 17:e0270429.
15 Uscinski J. *Conspiracy Theories: A Primer*. Rowman & Littlefield, 2020; Niskanen Center. Conspiracy beliefs are not increasing nor exclusive to the right. *Science of Politics* Episode 93; April 21, 2021. Transcript https://www.niskanencenter.org/conspiracy-beliefs-are-not-increasing-or-exclusive-to-the-right/
16 Dyer O. COVID-19: Unvaccinated face 11 times risk of death from delta variant, CDC data show. *BMJ* 2021;374:n2282; Frankovic K. Why won't Americans get vaccinated? YouGovAmerica.com. July 15, 2021. https://today.yougov.com/topics/politics/articles-reports/2021/07/15/why-wont-americans-get-vaccinated-poll-data
17 Jolley D, Douglas KM. The effects of anti-vaccine conspiracy theories on vaccination intentions. *PLoS ONE* 2014;9:e89177; Ripp T, Roer JP. Systematic review on the association of COVID-19-related conspiracy belief with infection-preventative behavior and vaccine willingness. *BMC Psychology* 2022;10:66.
18 YouGov. The Economist/YouGov Poll July 10–13, 2021. https://docs.cdn.yougov.com/w2zmwpzsq0/econTabReport.pdf
19 Oliver JE, Wood TJ. Medical conspiracy theories and health behaviors in the United States. *JAMA Internal Medicine* 2014;174:817–818.
20 Addley E. Study shows 60% of Britons believe in conspiracy theories. *The Guardian*. November 22, 2019. https://www.theguardian.com/society/2018/nov/23/study-shows-60-of-britons-believe-in-conspiracy-theories; YouGov. YouGov Cambridge Globalism

Project—Conspiracy theories. YouGov Poll. August 13–23, 2018. https://d25d2506sfb94s.cloudfront.net/cumulus_uploads/document/pk1qbgil4c/YGC%20Conspiracy%20Theories%20(all%20countries).pdf
21 Jolley D, Douglas KM, Marchlewska M, Cichocka A, Sutton RM. Examining the links between conspiracy theory beliefs and the EU "Brexit" referendum vote in the UK: Evidence from a two-wave survey. *Journal of Applied Social Psychology* 2021;00:1–7.
22 Cillizza C. 1 in 3 Americans believe the "Big Lie." CNN.com. June 21, 2021. https://www.cnn.com/2021/06/21/politics/biden-voter-fraud-big-lie-monmouth-poll/index.html.
23 Agiesta J. CNN poll: Most Americans think election results could lead to political violence in the coming years. CNN.com. March 12, 2021. https://www.cnn.com/2021/03/12/politics/cnn-poll-political-divisions/index.html; Hannon E. Poll finds nearly 40 percent of Republicans think political violence is justifiable and could be necessary. *Slate*. February 11, 2021. https://slate.com/news-and-politics/2021/02/aei-poll-40-percent-republicans-conservatives-political-violence.html
24 PRRI staff. Understanding QAnon's connection to American politics, religion, and media consumption. PRRI.org. May 27, 2021. https://www.prri.org/research/qanon-conspiracy-american-politics-report/
25 YouGov. The Economist/YouGov Poll. July 17–20, 2021. https://docs.cdn.yougov.com/1aaz80mjhy/econTabReport.pdf; Milman O, Harvey F. US is hotbed of climate change denial major global survey finds. *The Guardian*. May 8, 2019. https://www.theguardian.com/environment/2019/may/07/us-hotbed-climate-change-denial-international-poll; Ibbetson C. Where do people believe in conspiracy theories? YouGov. January 18, 2021. https://yougov.co.uk/topics/international/articles-reports/2021/01/18/global-where-believe-conspiracy-theories-true
26 Jolley D, Douglas KM. The social consequences of conspiracism: Exposure to conspiracy theories decreases intention to engage in politics and to reduce one's carbon footprint. *British Journal of Psychology* 2014;105:35–56; van Prooijen J-W, Krouwel APM. Pollet TV. Political extremism predicts belief in conspiracy theories. *Social Psychology and Personality Science* 2015;6:570–578; Jolley D, Douglas KM, Leite AC, Schrader T. Belief in conspiracy theories and intentions to engage in everyday crime. *British Journal of Psychology* 2019;58:534–549; Jolley D, Paterson JL. Pylons ablaze: Examining the role of 5G COVID-19 conspiracy beliefs and support for violence. *British Journal of Social Psychology* 2020;59:628–640; Vegetti F, Littvay L. Belief in conspiracy theories and attitudes toward political violence. *Italian Political Science Review* 2022;52:18–32; Jolley D, Marques MD, Cookson D. Shining a spotlight on the dangerous consequences of conspiracy theories. *Current Opinion in Psychology* 2022;47:101363; Toribio-Flórez D, Green R, Sutton RM, Douglas KM. Does belief in conspiracy theories affect interpersonal relationships? *Spanish Journal of Psychology* 2023;26(e9):1–8.
27 Winter J. Exclusive: FBI document warns conspiracy theories are a new domestic terrorism threat. Yahoo!News. August 1, 2019. https://news.yahoo.com/fbi-documents-conspiracy-theories-terrorism-160000507.html
28 Butter M. There's a conspiracy theory that the CIA invented the term "conspiracy theory"—here's why. *The Conversation*. March 16, 2020. https://theconversation.com/theres-a-conspiracy-theory-that-the-cia-invented-the-term-conspiracy-theory-heres-why-132117
29 Douglas KM, Sutton RM, Cichocka A. The psychology of conspiracy theories. *Current Directions in Psychological Science* 2017;26:538–542; Douglas KM, Uscinski JE, Sutton RM, et al. Understanding conspiracy theories. *Political Psychology* 2019;40(Suppl 1):3–35.
30 Brotherton R, French CC. Intention seekers: Conspiracist ideation and biased attributions of intentionality. *PLoS ONE* 2015;10:e0124125; Wagner-Egger P, Delouveé S, Gauvrit N, Dieguez S. Creationism and conspiracism share a common teleologic bias. *Current Biology* 2018;28:R867–R868.

31 Liekefett L, Christ O, Becker JC. Can conspiracy beliefs be beneficial? Longitudinal linkages between conspiracy theory beliefs, anxiety, uncertainty aversion, and existential threat. *Personality and Social Psychology Bulletin* 2023;49:167–179.
32 Lantian A, Muller D, Nurra C, Douglas KM. "I know things they don't know!" The role of need for uniqueness in belief in conspiracy theories. *Social Psychology* 2017;48:160–173.
33 Golec de Zavala A, Federico CM. Collective narcissism and growth of conspiracy thinking over the course of the 2016 United States presidential election: A longitudinal analysis. *European Journal of Social Psychology* 48:1011–1018.
34 Schaeffer K. Nearly three-in-ten Americans believe COVID-19 was made in a lab. Pew Research Center. April 8, 2020. https://www.pewresearch.org/fact-tank/2020/04/08/nearly-three-in-ten-americans-believe-covid-19-was-made-in-a-lab/; Uscinski JE, Enders AM, Klofstad C, et al. Why do people believe COVID-19 conspiracy theories? *Harvard Kennedy School Misinformation Review* 2020;1.
35 Pomfret J. The U.S.-China coronavirus blame game and conspiracies are getting dangerous. *The Washington Post*. March 17, 2020. https://www.washingtonpost.com/opinions/2020/03/17/us-china-coronavirus-blame-game-conspiracies-are-getting-dangerous/
36 Mackinnon A. Russia knows just who to blame for the coronavirus: America. Foreign Policy. February 14, 2020. https://foreignpolicy.com/2020/02/14/russia-blame-america-coronavirus-conspiracy-theories-disinformation/; O'Sullivan D. Exclusive. She's been falsely accused of starting the pandemic. Her life has been turned upside down. CNN.com. April 27, 2020. https://www.cnn.com/2020/04/27/tech/coronavirus-conspiracy-theory/index.html
37 Garrett RK, Weeks BE. Epistemic beliefs' role in promoting misperceptions and conspiracist ideation. *PLoS ONE* 2017;12(9):e0184733; Vranic A, Hromatko I, Tonkovic M. "I did my own research": Overconfidence, (dis)trust in science, and endorsement of conspiracy theories. *Frontiers in Psychology* 13:931865; Swami V, Voracek M, Stieger S, Tran US, Furnham A. Analytic thinking reduces belief in conspiracy theories. *Cognition* 2014;133:572–585; Yelbuz BE, Madan E, Alper S. Reflective thinking predicts lower conspiracy theory beliefs: A meta-analysis. *Judgment and Decision Making* 2022;17:720–744.
38 Miller JM, Saunders KL, Farhart CE. Conspiracy endorsement as motivated reasoning: The moderating roles of political knowledge and trust. *American Journal of Political Science* 2016;60:824–844; Smallpage SM, Enders AM, Uscinski JE. The partisan contours of conspiracy beliefs. *Research and Politics* 2017;October–December:1–7.
39 van der Linden S, Panagopoulos C, Azevedo F, Jost JT. The paranoid style in American politics revisited: An ideological asymmetry in conspiratorial thinking. *Political Psychology* 2012;42:23–51.
40 Enders A, Farhart C, Miller J, Uscinski J, Saunders K, Drochon H. Are Republicans and conservatives more likely to believe in conspiracy theories? *Political Behavior* 2023;45:2001–2004.
41 Pierre JM. Conspiracy theory belief: A sane response to an insane world? *Review of Philosophy and Psychology* 2023. https://doi.org/10.1007/s13164-023-00716-7
42 Kofta M, Soral W, Bilewicz M. What breeds conspiracy antisemitism? The role of political uncontrollability and uncertainty in the belief in Jewish conspiracy. *Journal of Personality and Social Psychology* 2020;118:900–918; van Prooijen J-W. An existential threat model of conspiracy theories. *European Psychology* 2020;25:16–25.
43 Jolley D, Meleady R, Douglas KM. Exposure to intergroup conspiracy theories promotes prejudice which spreads across groups. *British Journal of Psychology* 2020;111:17–35.
44 Ball K, Lawson W, Alim T. Medical mistrust, conspiracy beliefs and HIV-related behavior among African Americans. *Journal Psychology and Behavioral Science* 2013;1:1–7; Bogart LM, Thorburn S. Are HIV/AIDS conspiracy beliefs a barrier to HIV prevention among African Americans? *Journal of Acquired Immune Deficiency Syndrome* 2005;38:213–218.

45 Washington HA. *Medical Apartheid: The Dark History of Medical Experimentation on Black Americans from Colonial Times to the Present.* Doubleday, 2006.
46 Alper S. There are higher levels of conspiracy beliefs in more corrupt countries. *European Journal of Social Psychology* 2023;53:503–517.
47 McNeil DG. How much herd immunity is enough? *The New York Times.* December 24, 2020. https://www.nytimes.com/2020/12/24/health/herd-immunity-covid-coronavirus.html; Curet M. CDC did not say vaccines are failing or vaccinated people are superspreaders. *Politifact.* August 4, 2021. https://www.politifact.com/factchecks/2021/aug/04/instagram-posts/cdc-did-not-say-vaccines-are-failing-or-vaccinated/
48 Morisi D, Jost JT, Singh V. An asymmetrical "president-in-power" effect. *American Political Science Review* 2019;113:614–620.
49 Imhoff R, Zimmer F, Klein O, et al. Conspiracy mentality and political orientation across 26 countries. *Nature Human Behaviour* 2022;6:392–403.
50 Castanho Silva B, Vegetti F, Littvay L. The elite is up to something: Exploring the relation between populism and belief in conspiracy theories. *Swiss Political Science Review* 2017;23:423–443; Lewis P, Bosely S, Duncan P. Revealed: Populists far more likely to believe in conspiracy theories. *The Guardian.* May 1, 2019. https://www.theguardian.com/world/2019/may/01/revealed-populists-more-likely-believe-conspiracy-theories-vaccines; Eberl J-M, Huber RA, Greussing E. From populism to the "plandemic": Why populists believe in COVID-19 conspiracies. *Journal of Elections, Public Opinion and Parties* 2021;31:271–284; Stecula DA, Pickup M. How populism and conservative media fuel conspiracy beliefs about COVID-19 and what it means for COVID-19 behaviors. *Research and Politics* 2021; in press.
51 Uscinski JE, Enders AM, Seelig MI, et al. American politics in two dimensions: Partisan and ideological identities versus anti-establishment orientations. *American Journal of Political Science* 2021;65:877–895; Enders AM, Uscinski JE. The role of ant-establishment orientations during the Trump presidency. *The Forum.* 2021;19:47–76.
52 Washburn AN, Skitka LJ. Science denial across the political divide: Liberals and conservatives are similarly motivated to deny attitude-inconsistent science. *Social Psychology and Personality Science* 2018;9:972–980.
53 Gauchat G. Politicization of science in the public sphere: A study of public trust in the United States, 1974 to 2010. *American Sociological Review* 2012;77:167–187; McCright AM, Dentzman K, Charters M, Dietz T. The influence of political ideology on trust in science. *Environmental Research Letters* 2013:044029; Lewandowsky S, Oberauer K. Motivated rejection of science. *Current Directions in Psychological Science* 2016;25:217–222; Lewandowsky S, Oberauer K. Worldview-motivated rejection of science and the norms of science. *Cognition* 2021;215:104820; Lee JJ. Party polarization and trust in science: What about democrats? *Socius* 2021;7:1–12; Li N, Qian Y. Polarization of public trust in scientists between 1978 and 2018. *Politics and the Life Sciences* 2022;41:45–54.
54 Klein C, Clutton P, Dunn, AG. Pathways to conspiracy: The social and linguistic precursors of involvement in Reddit's conspiracy theory forum. *PLoS One* 2019;14:e0225098.
55 Landrum AR, Olshansky A, Richards O. Differential susceptibility to misleading flat earth arguments on Youtube. *Media Psychology* 2019;24:136–165.
56 Stempel C, Hargrove T, Stempel III GH. Media use, social structure, and belief in 9/11 conspiracy theories. *Journalism and Mass Communication Quarterly* 2007;84:353–372.
57 Lewis P, Bosely S, Duncan P. Revealed: Populists far more likely to believe in conspiracy theories. *The Guardian.* May 1, 2019. https://www.theguardian.com/world/2019/may/01/revealed-populists-more-likely-believe-conspiracy-theories-vaccines
58 Stecula DA, Pickup M. Social media, cognitive reflection, and conspiracy beliefs. *Frontiers in Political Science* 2021;3:647957.

59 Stecula DA, Pickup M. How populism and conservative media fuel conspiracy beliefs about COVID-19 and what it means for COVID-19 behaviors. *Research & Politics* 2021:8(1).
60 De Coninck D, Frissen T, Matthijs K, et al. Beliefs in conspiracy theories and misinformation about COVID-19: Comparative perspectives on the role of anxiety, depression and exposure to and trust in information sources. *Frontiers in Psychology* 2021;12:646394.
61 Uscinski JE, Enders AM. Don't blame social media for conspiracy theories—they would still flourish without it. *The Conversation*. June 18, 2020. https://theconversation.com/dont-blame-social-media-for-conspiracy-theories-they-would-still-flourish-without-it-138635; Enders AM, Uscinski JE, Seelig MI, et al. The relationship between social media use and beliefs in conspiracy theories and misinformation. *Political Behavior* 2023;45:781–804.
62 Niskanen Center. Conspiracy beliefs are not increasing nor exclusive to the right. *Science of Politics*, Episode 93; April 21, 2021. Transcript https://www.niskanencenter.org/conspiracy-beliefs-are-not-increasing-or-exclusive-to-the-right/
63 Holan AD. Ask PolitiFact: Are you sure Donald Trump didn't call the coronavirus a hoax? PolitiFact. October 8, 2020. https://www.politifact.com/article/2020/oct/08/ask-politifact-are-you-sure-donald-trump-didnt-cal/
64 Rosenblum NL, Muirhead R. *A Lot of People Are Saying: The New Conspiracism and the Assault on Democracy*. Princeton University Press, 2019.
65 Deer B. How the vaccine crisis was meant to make money. *BMJ* 2011;342:c5258.
66 Hussain A, Ali ZS, Ahmed M, Hussein S. The anti-vaccination movement: A regression in modern medicine. *Cureus* 2018;10:e2919.
67 Passantino J, Darcy O. Social media giants remove viral video with false coronavirus claims that Trump retweeted. CNN.com. July 28, 2020. https://www.cnn.com/2020/07/28/tech/facebook-youtube-coronavirus/index.html; America's Frontline Doctors SCOTUS press conference transcript. Rev.com. July 27, 2020. https://www.rev.com/blog/transcripts/americas-frontline-doctors-scotus-press-conference-transcript
68 Mackey T, Purushothaman V, Haupt M, Nali M, Li J. Application of unsupervised machine learning to identify and characterize hydroxychloroquine misinformation on Twitter. *Lancet Digital Health* 2021;3:e72–e75.
69 Nelson A. Anatomy of deceit: Team Trump deploys doctors with dubious qualifications to push fake cure for Covid-19. *The Washington Spectator*. September 20, 2020. https://washingtonspectator.org/anatomy-of-deceit/
70 Bergengruen V. How "America's frontline doctors" sold access to bogus COVID-19 treatments—and left patients in the lurch. *TIME*. August 26, 2021. https://time.com/6092368/americas-frontline-doctors-covid-19-misinformation/
71 Lee M. Network of right-wing health care providers in making millions off hydroxychloroquine and ivermectin, hacked data reveals. *The Intercept*. September 28, 2021. https://theintercept.com/2021/09/28/covid-telehealth-hydroxychloroquine-ivermectin-hacked/; Bergengruen V. How an online pharmacy sold millions worth of dubious COVID-19 drugs—while patients paid the price. *TIME*. October 13, 2021. https://time.com/6104407/ravkoo-pharmacy-ivermectin-covid-19-ppp-loan/
72 Sommer W. COVID-denying medical group implodes over founder's extravagant spending. *The Daily Beast*. November 14, 2022. https://www.thedailybeast.com/covid-misinformation-group-americas-frontline-doctors-implodes-over-dr-simone-golds-extravagant-spending
73 Merlan A. On the dark and dangerous underbelly of climate conspiracy theories. *LitHub*. September 17, 2019. https://lithub.com/on-the-dark-and-dangerous-underbelly-of-climate-conspiracy-theories/
74 Cook J, Supran G, Lewandowsky S, Oreskes N, Maibach E. America misled: How the fossil fuel industry deliberately misled Americans about climate change. George Mason University Center for Climate Change Communication, 2019. https://www.climatechangecommunication.org/america-misled/

75 Geary J. The dark money of climate change. *ESSAI* 2019;17:17. https://dc.cod.edu/essai/vol17/iss1/17
76 Mangan D. Trump rioter, QAnon supporter Douglas Austin Jensen thought he invaded White House, not Capitol, video shows. CNBC.com. July 12, 2021. https://www.cnbc.com/2021/07/13/trump-rioter-douglas-austin-jensen-thought-he-invaded-white-house-not-capitol.html
77 Hsu S. QAnon "poster boy" for Capitol riot sent back to jail after violating court order to stay off internet. *The Washington Post*. September 2, 2021. https://www.washingtonpost.com/local/legal-issues/douglas-jensen-jailed-qanon-addiction/2021/09/02/50ee9628-0c08-11ec-aea1-42a8138f132a_story.html
78 Elfrink T. He wore a QAnon shirt while chasing police on Jan. 6. Now he says he was deceived by "a pack of lies." *The Washington Post*. June 8, 2021. https://www.washingtonpost.com/nation/2021/06/08/douglas-jensen-qanon-conspiracy/
79 Pew Research Center. 5 facts about the QAnon conspiracy theories. PewResearch.org. November 16. 2020. https://www.pewresearch.org/fact-tank/2020/11/16/5-facts-about-the-qanon-conspiracy-theories/; Rose J. Even if it's "bonkers," poll finds many believe QAnon and other conspiracy theories. NPR.org. December 30, 2020. https://www.npr.org/2020/12/30/951095644/even-if-its-bonkers-poll-finds-many-believe-qanon-and-other-conspiracy-theories
80 Chang A. We analyzed every QAnon post on Reddit. Here's who QAnon supporters actually are. *Vox*. August 8, 2018. https://www.vox.com/2018/8/8/17657800/qanon-reddit-conspiracy-data
81 LaFrance A. The prophesies of Q. *The Atlantic*. June 2020. https://www.theatlantic.com/magazine/archive/2020/06/qanon-nothing-can-stop-what-is-coming/610567/; Argentino M-A. The Church of QAnon: Will conspiracy theories form the basis of a new religious movement? *The Conversation*. May 18, 2020. https://theconversation.com/the-church-of-qanon-will-conspiracy-theories-form-the-basis-of-a-new-religious-movement-137859
82 Pierre JM. Down the conspiracy theory rabbit hole: How does one become a follower of QAnon? In: Miller MK, ed. *The Social Science of QAnon*. Cambridge University Press, 2023, 17–32.
83 Edelman G. QAnon supporters aren't quite who you think they are. *Wired*. October 6, 2020. https://www.wired.com/story/qanon-supporters-arent-quite-who-you-think-they-are/; Schaffner B. QAnon and conspiracy beliefs. September 18–20, 2020. https://www.isdglobal.org/wp-content/uploads/2020/10/qanon-and-conspiracy-beliefs.pdf
84 Carter B. Trump, addressing far-right QAnon conspiracy, offers praise for its followers. NPR.com. August 19, 2020. https://www.npr.org/2020/08/19/904055593/trump-addressing-far-right-qanon-conspiracy-offers-praise-for-its-followers
85 Scott D. Trump refuses to say the QAnon conspiracy theory is false. *Vox*. October 15, 2020. https://www.vox.com/2020/10/15/21518697/donald-trump-town-hall-what-is-qanon-conspiracy-theory
86 McClatchey E. Iowan facing Jan, 6 charges claims he "got taken" by QAnon, is granted pretrial release. *Little Village Magazine*. July 15, 2021. https://littlevillagemag.com/iowan-facing-jan-6-charges-claims-he-got-taken-by-qanon-is-granted-pretrial-release/

Chapter 7

1 Brotherton R, French CC, Pickering. Measuring belief in conspiracy theories: The generic conspiracist beliefs scale. *Frontiers in Psychology* 2013;4:279.
2 Frankfurt H. On Bullshit. *Raritan Quarterly Review* 1986;6:81–100.
3 Petrocelli JV. Antecedents of bullshitting. *Journal of Experimental Social Psychology* 2018;76:249–258.

4. Pennycook G, Cheyne JA, Barr N, Koehler DJ, Fugelsang JA. On the reception and detection of pseudo-profound bullshit. *Judgment and Decision Making* 2015;10:549–563.
5. Frankfurt H. *On Bullshit*. Princeton University Press, 2005.
6. MacKenzie A, Bhatt I. Lies, bullshit and fake news: Some epistemological concerns. *Postdigital Science and Education* 2020;2:9–13.
7. Ig Nobel Prize winners. https://www.improbable.com/2021-ceremony/winners/
8. Hart J, Graether M. Something's going on here: Psychological predictors of belief in conspiracy theories. *Journal of Individual Differences* 2018;39:229–237; Čavojová V, Secară EC, Jurkovič M, Šrol J. Reception and willingness to share pseudo-profound bullshit and their relation to other epistemically suspect beliefs and cognitive ability in Slovakia and Romania. *Applied Cognitive Psychology* 2019;33:299–311.
9. Littrell S, Risko EF, Fugelsang JA. "You can't bullshit a bullshitter" (or can you?): Bullshitting frequency predicts receptivity to various types of misleading information. *British Journal of Social Psychology* 2021;60:1484–1505.
10. Littrell S, Fugelsang JA. Bullshit blind spots: The roles of miscalibration and information processing in bullshit detection. *Thinking & Reasoning* 2023. https://doi.org/10.1080/13546783.2023.2189163
11. Pennycook G, Cheyne JA, Barr N, Koehler DJ, Fugelsang JA. It's still bullshit: Reply to Dalton (2016). *Judgment and Decision Making* 2016;11:123–125.
12. Pennycook G, Rand DG. Who falls for fake news? The roles of bullshit receptivity, overclaiming, familiarity, and analytic thinking. *Journal of Personality* 2020;88:185–200.
13. Bainbridge TF, Quinlan JA, Mar RA, Smillie LD. Openness/intellect and susceptibility to pseudo-profound bullshit: A replication and extension. *European Journal of Personality* 2018;33:72–88; Čavojová V, Secară EC, Jurkovič M, Šrol J. Reception and willingness to share pseudo-profound bullshit and their relation to other epistemically suspect beliefs and cognitive ability in Slovakia and Romania. *Applied Cognitive Psychology* 2019;33:299–311.
14. Chotiner I. Deepak Chopra has never been sick. *The New Yorker*. October 17, 2019. https://www.newyorker.com/culture/q-and-a/deepak-chopra-has-never-been-sick
15. Tompkins P. New age supersage. *Time*. November 14, 2008. https://time.com/archive/6893750/new-age-supersage/
16. Chopra Global. Tonia O'Connor appointed CEO of Deepak Chopra's next generation wellbeing company Chopra Global. *PR Newswire*. June 25, 2019. https://www.prnewswire.com/news-releases/tonia-oconnor-appointed-ceo-of-deepak-chopras-next-generation-well-being-company-chopra-global-300874174.html
17. Chopra D. *The Seven Spiritual Laws of Success*. Amber-Allen Publishing, 1994.
18. Goldhill O. We asked Deepak Chopra, the guru of sayings that mean nothing, to fact-check his own tweets. *Quartz*. March 5, 2017. https://qz.com/917820/we-asked-deepak-chopra-the-guru-of-sayings-that-mean-nothing-to-fact-check-his-own-tweets/
19. Chopra D, Redwood D. Interviews with people who make a difference: Quantum healing. Healthy.net. 1995. http://184.154.57.4/scr/interview.aspx?Id=167
20. Green E. Understanding Deepak Chopra's "biofields." *The Atlantic*. October 4, 2013. https://www.theatlantic.com/health/archive/2013/10/understanding-deepak-chopras-biofields/280248/
21. Sperber D. The guru effect. *Review of Philosophy and Psychology* 2010;1:583–592.
22. Evans A, Sleegers W, Mlakar Z. Individual differences in receptivity to scientific bullshit. *Judgment and Decision Making* 2020;15:401–412.
23. Vranic A, Hromatko I, Tonkovic M. "I did my own research": Overconfidence, (dis)trust in science, and endorsement of conspiracy theories. *Frontiers in Psychology* 2022;13:931865.
24. Boudry M, Blancke S, Pigliucci M. What makes weird beliefs thrive? The epidemiology of pseudoscience. *Philosophical Psychology* 2015;28: 1177–1198.

25 O'Mathuna DP. Evidence-based practice and reviews of therapeutic touch. *Journal of Nursing Scholarship* 2000;32:279–285.
26 Rosa L, Rosa E, Sarner L, Barrett S. A close look at therapeutic touch. *JAMA* 1998;279:1005–1010.
27 Rosa E. TT and me. *Jr. Skeptic* 1998;1:3–5. https://shop.skeptic.com/junior-skeptic-1-emily-rosa-and-friend
28 Chopra D, Barsotti T, Mills PJ. Surprising findings about biofield healing. *Medium.* July 13, 2020. https://deepakchopra.medium.com/surprising-findings-about-biofield-healing-568c334938eb
29 Jain S, Pavlik D, Distefan J, et al. Complementary medicine for fatigue and cortisol variability in breast cancer survivors. *Cancer* 2011;118:777–787.
30 Garrett B, Riou M. A rapid evidence assessment of recent therapeutic touch research. *Nursing Open* 2020;8:8318–2330.
31 Ward C, Voas D. The emergence of conspirituality. *Journal of Contemporary Religion* 2011;26:103–121.
32 Gligorić V, da Silva MM, Eker S, et al. The usual suspects: How psychological motives and thinking styles predict endorsement of well-known and COVID-19 conspiracy beliefs. *Applied Cognitive Psychology* 2021;35:1171–1181.
33 Aubrey S. "Playing with fire": The curious marriage of QAnon and wellness. *Sydney Morning Herald.* September 27, 2020. https://www.smh.com.au/lifestyle/health-and-wellness/playing-with-fire-the-curious-marriage-of-qanon-and-wellness-20200924-p55yu7.html; Love S. "Conspirituality" explains why the wellness world fell for QAnon. *Vice.* December 16, 2020. https://www.vice.com/en/article/93wq73/conspirituality-explains-why-the-wellness-world-fell-for-qanon; Wiseman E. The dark side of wellness: The overlap between spiritual thinking and far-right conspiracies. *The Guardian.* October 17, 2021. https://www.theguardian.com/lifeandstyle/2021/oct/17/eva-wiseman-conspirituality-the-dark-side-of-wellness-how-it-all-got-so-toxic; Pierre JM. Down the conspiracy theory rabbit hole: How does one become a follower of QAnon? In: Miller MK, ed. *The Social Science of QAnon.* Cambridge University Press, 2023, 17–32.
34 Finger B. Do you love "wise-sounding" quotes? Surprise! You're probably dumb. *Jezebel.* December 4, 2015. https://jezebel.com/do-you-love-wise-sounding-quotes-surprise-youre-proba-1746220367
35 Kohnhorst A. People who post inspirational Facebook quotes are morons, according to science. *Maxim.* December 4, 2015. https://www.maxim.com/gear/profound-quotes-are-bullshit-science-2015-12/
36 Sokal AD. Transgressing the boundaries: Toward a transformative hermeneutics of quantum gravity. *Social Text* 1996;14:217–252.
37 Sokal AD. A physicist experiments with cultural studies. *Lingua Franca* 1996;6:62–64.
38 Counterbalance Foundation. Postmodernism. https://counterbalance.org/gengloss/postmbody.html
39 Dennett DC. Dennett on Wieseltier v. Pinker in the New Republic. *Edge.* September 10, 2013. https://www.edge.org/conversation/dennett-on-wieseltier-v-pinker-in-the-new-republic
40 Gligorić V, Vilotejević A. "Who said it?" How contextual information influences perceived profundity of meaningful quotes and pseudo-profound bullshit. *Applied Cognitive Psychology* 2020;34:535–542; Ilic S, Dmanjanovic K. The effect of source credibility on bullshit receptivity. *Applied Cognitive Psychology* 2021;35:1193–1205.
41 Ruark J. Bait and switch. *Chronicle of Higher Education.* January 1, 2017. http://chronicle.com/article/bait-and-switch
42 Sokal AD. Pseudoscience and postmodernism: Antagonists or fellow-travelers? In: Fagan G, ed. *Archaeological Fantasies: How Pseudoarchaeology Misrepresents the Past and Misleads the Public.* Routledge, 2006.

43 Boyd B. Brian Boyd: Bullshit can be more dangerous than lies in politics. *The Irish Times.* June 3, 2106. https://www.irishtimes.com/opinion/brian-boyd-bullshit-can-be-more-dangerous-than-lies-in-politics-1.2670629; Foroughi H, Fotaki M, Gabriel Y. Why leaders who bullshit are more dangerous than those who lie. *The Conversation.* November 12, 2019. https://theconversation.com/why-leaders-who-bullshit-are-more-dangerous-than-those-who-lie-125109
44 Pierre JM. Assessing malingered auditory verbal hallucinations in forensic and clinical settings. *Journal of the American Academy of Psychiatry and the Law* 2019;47:448–456.
45 Pierre JM, Wirshing DA, Wirshing WC. "Iatrogenic malingering" in VA substance abuse treatment. *Psychiatric Services* 2003;54:253–254.
46 Spicer A. Playing the bullshit game: How empty and misleading communication takes over organizations. *Organization Theory* 2020;1:1–26.
47 Funk C, Hefferon M, Kennedy B, Johnson C. Trust and mistrust in American' views of scientific experts. Pew Research Center. August 2, 2019. https://www.pewresearch.org/science/2019/08/02/trust-and-mistrust-in-americans-views-of-scientific-experts/
48 Holon AD. All politicians lie. Some lie more than others. *The New York Times.* December 11, 2015. https://www.nytimes.com/2015/12/13/opinion/campaign-stops/all-politicians-lie-some-lie-more-than-others.html
49 Sterling J, Jost JT, Pennycook G. Are neoliberals more susceptible to bullshit? *Judgment and Decision Making* 2016;11:352–360.
50 Pfattheicher S, Schinlder S. Misperceiving bullshit as profound is associated with favorable views of Cruz, Rubio, Trump and conservatism. *PLoS ONE* 11(4):e0153419.
51 Nilsson A, Erlandsson A, Västfjäll D. The complex relation between receptivity to pseudo-profound bullshit and political ideology. *Personality and Social Psychology Bulletin* 2019;45:1440–1454.
52 Petrocelli JV. Politically oriented bullshit detection: Attitudinal conditional bullshit receptivity and bullshit sensitivity. *Group Processes and Intergroup Relations* 2022;25:1635–1652.
53 Littrell S, Meyers EA, Fugelsang JA. Not all bullshit pondered is tossed: Reflection decreases receptivity to some types of misleading information but not others. *Applied Cognitive Psychology* 2024;38:e4154.
54 Bergstrom C, West J. FAQ: Frequently asked questions. Callingbullshit.org. https://www.callingbullshit.org/FAQ.html
55 Bergstrom C, West J. *Calling Bullshit: The Art of Skepticism in a Data-Driven World.* Random House, 2020.

Chapter 8

1 Pierre J. The psychological needs that QAnon feeds. Psych unseen: Brain, behavior, and belief. *Psychology Today.* August 12, 2020. https://www.psychologytoday.com/us/blog/psych-unseen/202008/the-psychological-needs-qanon-feeds; Pierre J. How far down the QAnon rabbit hole did your loved one fall? Psych unseen: Brain, behavior, and belief. *Psychology Today.* August 21, 2020. https://www.psychologytoday.com/us/blog/psych-unseen/202008/how-far-down-the-qanon-rabbit-hole-did-your-loved-one-fall; Pierre J. 4 keys to help someone climb out of the QAnon rabbit hole. Psych unseen: Brain, behavior, and belief. *Psychology Today.* September 1, 2020. https://www.psychologytoday.com/us/blog/psych-unseen/202009/4-keys-help-someone-climb-out-the-qanon-rabbit-hole
2 Pierre JM. Down the conspiracy theory rabbit hole: How does one become a follower of QAnon? In: Miller MK, ed. *The Social Science of QAnon.* Cambridge University Press, 2023, 17–32; Carrier A. "This crap means more to him than my life": When QAnon invades American homes. *Politico.* February 19, 2021. https://www.politico.com/news/

magazine/2021/02/19/qanon-conspiracy-theory-family-members-reddit-forum-469485; Niz E. QAnon is ruining families, but some moms and kids are fighting back. *Yahoo!life.* February 11, 2021. https://www.yahoo.com/lifestyle/qanon-ruining-families-moms-kids-222743014.html; Watt CS. The QAnon orphans: People who have lost loved ones to conspiracy theories. *The Guardian.* September 23, 2020. https://www.theguardian.com/us-news/2020/sep/23/qanon-conspiracy-theories-loved-ones; Hawkins E. Dr. Joseph Pierre: UCLA psychiatrist takes on QAnon & "rescuing" loved ones. *Heavy.* September 9, 2020. https://heavy.com/news/2020/09/joseph-pierre-qanon/; Minutaglio R. My boyfriend reads QAnon theories. I still love him—but I'm worried. *Esquire.* August 24, 2018. https://www.esquire.com/news-politics/a22664244/qanon-boyfriend-conspiracy-theorist-my-partner-deep-state/
3   Rauch J. Bipolar disorder: Is America divided? Brookings.edu. January 1, 2005. https://www.brookings.edu/articles/bipolar-disorder-is-america-divided; Hendricks S. Why is the United States so divided? Simple, it was never united at all. BigThink.com. December 20, 2016. https://bigthink.com/the-present/why-is-the-united-states-so-divided-simple-it-was-never-united-at-all; Moss DA. *Democracy: A Case Study.* Belknap Press of Harvard University Press, 2017.
4   Pew Research Center. The partisan divide of political values grows even wider. October 5, 2017. https://www.pewresearch.org/politics/2017/10/05/the-partisan-divide-on-political-values-grows-even-wider/
5   Baker JA, III. John McCain and the dying art of political compromise. *Wall Street Journal.* August 29, 2018. https://www.wsj.com/articles/john-mccain-and-the-dying-art-of-political-compromise-1535581611; Montanaro D. McCain's death marks the near-extinction of bipartisanship. NPR.org. August 30, 2018. https://www.npr.org/2018/08/30/642409720/mccains-death-marks-the-near-extinction-of-bipartisanship
6   Shephard A. Bipartisanship is dead, and that's great news for Joe Biden. *The New Republic.* January 21, 2022. https://newrepublic.com/article/165093/bipartisanship-dead-biden-democrats-obama
7   Folkenflik D. Building bipartisanship? Not Limbaugh's problem. NPR.org. January 25, 2007. https://www.npr.org/templates/story/story.php?storyId=7018083
8   Iyengar S, Sood G, Lelkes Y. Affect: Not ideology: A social identity perspective on polarization. *Public Opinion Quarterly* 2012;76:405–431.
9   Finkel EJ, Bail CA, Cikara M, et al. Political sectarianism in America. *Science* 2020;370:533–536.
10 YouGov. The Economist/YouGov Poll. September 13–15, 2020. https://docs.cdn.yougov.com/t0hi1tcqs5/econTabReport.pdf; Opzoomer I. America speaks: What do they think about cross-party marriages? YouGovAmerica.com. September 24, 2020. https://today.yougov.com/topics/lifestyle/articles-reports/2020/09/24/america-speaks-what-do-they-think-about-cross-part
11 Mason L. Ideologues without issues: The polarizing consequences of ideological identities. *Public Opinion Quarterly* 2018,82: 280–301.
12 Landy JF, Rottman J, Batres C, Leimgruber KL. Disgusting Democrats and repulsive Republicans: Members of political outgroups are considered physically gross. *Personality and Social Psychology Bulletin* 2023;49:361–375.
13 Iyengar S, Westwood SJ. Fear and loathing across party lines: New evidence on group polarization. *American Journal of Political Science* 2015;59:690–707.
14 Flynn DJ, Nyhan B, Reifler J. The nature and origins of misperceptions: Understanding false and unsupported beliefs about politics. *Advances in Political Psychology* 2017;38:127–150.
15 Stanovich KE, West RF, Toplak ME. Myside bias, rational thinking, and intelligence. *Current Directions in Psychological Science* 2013;22:259–264; Stanovich KE. *The Bias That Divides Us: The Science and Politics of Myside Thinking.* MIT Press, 2021.

16 Jenke L. Affective polarization and misinformation belief. *Political Behavior* 2023. https://doi.org/10.1007/s11109-022-09851-w
17 Appiah KA. People don't vote for what they want. They vote for who they are. *The Washington Post*. August 30, 2018. https://www.washingtonpost.com/outlook/people-dont-vote-for-want-they-want-they-vote-for-who-they-are/2018/08/30/fb5b7e44-abd7-11e8-8a0c-70b618c98d3c_story.html
18 Sachdev R. From the White House to ancient Athens: Hypocrisy is not match for partisanship. *The Conversation*. December 11, 2020. https://theconversation.com/from-the-white-house-to-ancient-athens-hypocrisy-is-no-match-for-partisanship-148504; Feldmann L. How political tribalism is leading to more political hypocrisy. *The Christian Science Monitor*. January 10, 2020. https://www.csmonitor.com/USA/Politics/2020/0110/How-political-tribalism-is-leading-to-more-political-hypocrisy
19 Moody C. Trump in '04: "I probably identify more as Democrat." CNN.com. July 22, 2015. https://www.cnn.com/2015/07/21/politics/donald-trump-election-democrat/index.html; Meet the Press. Trump in 1999: "I am very pro-choice." NBC News. July 8, 2018. https://www.nbcnews.com/meet-the-press/video/trump-in-1999-i-am-very-pro-choice-480297539914.
20 Sonnad N. These are Americans' favorite insults, by political affiliation. *Quartz*. November 4, 2014. https://qz.com/291533/this-is-how-liberals-and-conservatives-insult-each-other/
21 Pierre JM. No, the problem with America isn't "mass psychosis." Psych unseen: Brain, behavior, and belief. *Psychology Today*. January 14, 2020. https://www.psychologytoday.com/us/blog/psych-unseen/202201/no-the-problem-america-isnt-mass-psychosis; Pierre JM. Does "mass formation psychosis" really exist? Psych unseen: Brain, behavior, and belief. *Psychology Today*. January 14, 2020. https://www.psychologytoday.com/us/blog/psych-unseen/202201/does-mass-formation-psychosis-really-exist
22 Levin B. Republican lawmaker tells Alexandria Ocasio-Cortez to "relax" about video he made depicting her murder. *Vanity Fair*. November 9, 2021. https://www.vanityfair.com/news/2021/11/paul-gosar-alexandria-ocasio-cortez-video
23 Orhan YE. The relationship between affective polarization and democratic backsliding: Comparative evidence. *Democratization* 2022;29:714–735.
24 McCoy J, Press B. What happens when democracies become perniciously polarized? Carnegie Endowment for International Peace. January 22, 2022. https://carnegieendowment.org/2022/01/18/what-happens-when-democracies-become-perniciously-polarized-pub-86190
25 Kalmoe NP. Fueling the fire: Violent metaphors, trait aggression, and support for political violence. *Political Communications* 2014;31:545–563; Kalmoe NP, Gubler JR, Wood DA. Toward conflict or compromise? How violent metaphors polarize partisan issue attitudes. *Political Communications* 2018;35:333–352.
26 Mason L, Kalmoe NP. What you need to know about how many Americans condone political violence—and why. *The Washington Post*. January 11, 2021. https://www.washingtonpost.com/politics/2021/01/11/what-you-need-know-about-how-many-americans-condone-political-violence-why/
27 Kornfield M, Alfaro M. 1 in 3 Americans say political violence against government can be justified, citing fears of political schism, pandemic. *The Washington Post*. January 1, 2022. https://www.washingtonpost.com/politics/2022/01/01/1-3-americans-say-violence-against-government-can-be-justified-citing-fears-political-schism-pandemic/; Thomson-DeVeaux A. Why many Americans might be increasingly accepting of political violence. FiveThirtyEight.com. January 6, 2022. https://fivethirtyeight.com/features/why-many-americans-might-be-increasingly-accepting-of-political-violence/
28 Conroy M, Bacon P, Jr. There's a huge gap in how Republicans and Democrats see discrimination. Fivethirtyeight.com. June 17, 2020. https://fivethirtyeight.com/features/

theres-still-a-huge-partisan-gap-in-how-americans-see-discrimination/; Tesler M. Republicans and Democrats agree on the protests but not why people are protesting. Fivethirtyeight.com. June 17, 2020. https://fivethirtyeight.com/features/republicans-and-democrats-increasingly-agree-on-the-protests-but-not-why-people-are-protesting/

29. Mason L. "I respectfully disagree": The differential effects of partisan sorting on social and issue polarization. *American Journal of Political Science* 2015;59:128–145.

30. Clark CJ, Liu BS, Winegard BM, Ditto P. Tribalism is human nature. *Current Directions in Psychological Science* 2019;28:587–592.

31. Greenwald AG, McGhee D, Schwartz JLK. Measuring individual differences in implicit cognition: The implicit association test. *Journal of Personality and Social Psychology* 1998;74:1464–1480.

32. Nosek BA, Smyth FL, Hansen JJ, et al. Pervasiveness and correlates of implicit attitudes and stereotypes. *European Review of Social Psychology* 2007;18:36–88.

33. Blanton H, Jaccard J, Klick J, Mellers B, Mitchell G, Tetlock. Strong claims and weak evidence: Reassessing the predictive validity of the IAT. *Journal of Applied Psychology* 2009;94:567–582; Arkes HR, Tetlock PE. Attributions of implicit prejudice, or "would Jesse Jackson 'fail' the implicit association test?" *Psychological Inquiry* 2004;15:257–278; Mitchell G, Tetlock PE. Popularity as a poor proxy for utility. In: Lilienfeld SO, Waldman ID, eds. *Psychological Science under Scrutiny: Recent Challenges and Proposed Solutions.* Wiley, 2017; Singal J. Psychology's favorite tool for measuring racism isn't up to the job. *NewYorkMagazine.*January11,2017.https://www.thecut.com/2017/01/psychologys-racism-measuring-tool-isnt-up-to-the-job.html; Schimmack U. The implicit association test: A method in search of a construct. *Perspectives on Psychological Science* 2021:396–414.

34. Greenwald AG, Poehlman TA, Uhlmann EL, Banaji MR. Understanding the using the Implicit Association Test: III. Meta-analysis of predictive validity. *Journal of Personality and Social Psychology* 2009;97:17–41; Kurdi B, Seitchik AE, et al. Relationship between the implicit association test and intergroup behavior: A meta-analysis. *American Psychologist* 2019;74:569–586.

35. Morin R. Exploring racial bias among biracial and single-race adults: The IAT. Pew Research Center. August 19, 2015. https://www.pewresearch.org/social-trends/2015/08/19/exploring-racial-bias-among-biracial-and-single-race-adults-the-iat/

36. Payne KB, Hannay JW. Implicit bias reflects systemic racism. *Trends in Cognitive Sciences* 2021;25:927–936.

37. Payne BK, Vuletich HA, Lundberg KB. The bias of crowds: How implicit bias bridges personal and systemic prejudice. *Psychological Inquiry* 2017;28:233–248

38. Jost JT. A decade of system justification theory: Accumulated evidence of conscious and unconscious bolstering of the status quo. *Political Psychology* 2004;25:881–919; Jost JT. The IAT is dead, long live the IAT: Context-sensitive measures of implicit attitudes are indispensable to social and political psychology. *Current Directions in Psychological Science* 2019;28:10–19; Jost JT. A quarter century of system justification theory: questions, answers, criticisms, and societal applications. *British Journal of Social Psychology* 2019;58:262–314.

39. Singal J. Psychology's favorite tool for measuring racism isn't up to the job. *New York Magazine.* January 11, 2017. https://www.thecut.com/2017/01/psychologys-racism-measuring-tool-isnt-up-to-the-job.html

40. Essien I, Calanchini J, Degner J. Moderators of intergroup evaluation in disadvantaged groups: A comprehensive test of predictions from system justification theory. *Journal of Personality and Social Psychology: Interpersonal Relations and Group Processes* 2021;120:1204–1230.

41. Knowles ED, Lowery BS, Chow RM, Unzueta MM. Deny, distance, or dismantle? How White Americans manage a privileged identity. *Perspectives on Psychological Science* 2014;9:594–609.

42 Jaimungal C. How does the "Black Lives Matter" slogan resonate with America? YouGovAmerica. June 22, 2020. https://today.yougov.com/topics/politics/articles-reports/2020/06/22/black-lives-matter-slogan-poll
43 Gibbs N. How Donald Trump lost by winning. *TIME*. March 1, 2019. https://time.com/5542123/donald-trump-michael-cohen-2016-campaign/
44 Bouise J. How Trump happened. *Slate*. March 13, 2016. http://www.slate.com/articles/news_and_politics/cover_story/2016/03/how_donald_trump_happened_racism_against_barack_obama.html; Thompson D. Donald Trump and the twilight of white America. *The Atlantic*. May 13, 2016. https://www.theatlantic.com/politics/archive/2016/05/donald-trump-and-the-twilight-of-white-america/482655/
45 Resnick B. White fear of demographic change is a powerful psychological force. *Vox*. January 28, 2017. https://www.vox.com/science-and-health/2017/1/26/14340542/white-fear-trump-psychology-minority-majority; Serwer A. The nationalist's delusion. *The Atlantic*. November 20, 2017. https://www.theatlantic.com/politics/archive/2017/11/the-nationalists-delusion/546356/; Chokshi N. Trump voters driven by fear of losing status, not economic anxiety, study finds. *The New York Times*. April 24, 2018. https://www.nytimes.com/2018/04/24/us/politics/trump-economic-anxiety.html;
46 Craig MA, Richeson JA. On the precipice of a "majority-minority" America: Perceived status threat from the racial demographic shift affects white Americans' political ideology. *Psychological Science* 2014;25:1189–1197.
47 Mutz DC. Status threat, not economic hardship, explains the 2016 presidential vote. *PNAS* 2018;115:E4330–E4339.
48 Jardina A. In-group love and out-group hate: White racial attitudes in contemporary U.S. election. *Political Behavior* 2020;43:1535–1559.
49 PRRI. New PRRI/The Atlantic survey analysis finds cultural displacement—not economic hardship—more predictive of white working-class support for Trump. PRRI.org. May 9, 2017. https://www.prri.org/press-release/white-working-class-attitudes-economy-trade-immigration-election-donald-trump/
50 Chicago Project on Security and Threats. American face of insurrection: Analysis of individuals charged for storming the US Capitol on January 6, 2021. January 2, 2022. https://d3qi0qp55mx5f5.cloudfront.net/cpost/i/docs/Pape_-_American_Face_of_Insurrection_(2022-01-05).pdf?mtime=1654548769; Pape RA. Opinion: What an analysis of 377 Americans arrested or charged in the Capitol insurrection tells us. *The Washington Post*. April 6, 2021. https://www.washingtonpost.com/opinions/2021/04/06/capitol-insurrection-arrests-cpost-analysis/#
51 Alesina A, Miano A, Stantcheva S. The polarization of reality. *AEA Papers and Proceedings* 2020;11:324–328.
52 Tapper J. Did Obama say, "If you've got a business, you didn't build that"? abcnews.go.com. July 16, 2012. https://abcnews.go.com/blogs/politics/2012/07/did-obama-say-if-youve-got-a-business-you-didnt-build-that
53 Jost JT. Ideological asymmetries and the essence of political psychology. *Political Psychology* 2017;38:167–208.
54 Zmigrod L, Rentfrow PJ, Robbins TW. The partisan mind: Is extreme political partisanship related to cognitive flexibility? *Journal of Experimental Psychology: General* 2020;149:407–418; Zmigrod L. The role of cognitive rigidity in political ideologies: Theory, evidence, and future directions. *Current Opinion in Behavioral Sciences* 2020;34:34–39.
55 Jost JT, Haperin E, Laurin K. Editorial overview: Five observations about tradition and progress in the scientific study of political ideologies. *Current Opinion in Behavioral Sciences* 2020;34:iii–vii.
56 Azarian B. A neuroscientist explains what may be wrong with Trump supporters' brains. Rawstory.com. August 4, 2016. https://www.rawstory.com/2016/08/a-neuroscientist-explains-what-may-be-wrong-with-trump-supporters-brains/

57 Ditto PH, Liu BS, Clark CJ, et al. At least bias is bipartisan: A meta-analytic comparison of partisan bias in liberals and conservatives. *Perspectives on Psychological Science* 2019;14:273–291; Barron J, Jost JT. False equivalence: Are liberals and conservatives in the United States equally biased? *Perspectives on Psychological Science* 2019;14:292–303; Ditto PH, Clark CJ, Liu BS, et al. Partisan bias and its discontents. *Perspectives on Psychological Science* 2019;14:304–216.
58 Brandt MJ, Reyna C, Chambers JR, Crawford JT, Wetherell G. The ideological-conflict hypothesis: intolerance among both liberals and conservatives. *Current Directions in Psychological Science* 2014;23:27–34.
59 Fisher M, Keil FC. The binary bias: A systematic distortion in the integration of information. *Psychological Science* 2018;29:1846–1858.
60 Pew Research Center. Beyond red vs. blue: The political typology. November 2021. https://www.pewresearch.org/politics/2021/11/09/beyond-red-vs-blue-the-political-typology-2/
61 Osei-Opare N. When it comes to America's race issues, Russia is a bogeyman. Foreignpolicy.com. July 6, 2020. https://foreignpolicy.com/2020/07/06/when-it-comes-to-americas-race-issues-russia-is-a-bogeyman/
62 Hawkins S, Yudkin D, Juan-Torres M, Dixon T. Hidden tribes: A study of America's polarized landscape. More in Common. 2018. https://hiddentribes.us/media/qfpekz4g/hidden_tribes_report.pdf
63 Hidden Tribes. COVID-19: Polarization and the pandemic. More in Common/YouGov. April 3, 2020. https://www.moreincommon.com/media/z4fdmdpa/hidden-tribes_covid-19-polarization-and-the-pandemic-4-3-20.pdf
64 Kubin E, Puryear C, Schein C, Gray K. Personal experiences bridge moral and political divides better than facts. *PNAS* 2021;118:e2008389918.
65 Stern K. American's aren't as divided as you think. *Politico*. November 19, 2017. https://www.politico.com/magazine/story/2017/11/19/americans-divided-politics-unity-liberal-bubble-215843/; Stern K. *Republican Like Me: How I Left the Liberal Bubble and Learned to Love the Right*. Harper Collins, 2017.
66 Griffin JH. *Black Like Me*. Signet Books, 1960.

Chapter 9

1 Festinger L. *A Theory of Cognitive Dissonance*. Stanford University Press, 1957.
2 Festinger L, Riecken H, Schachter S. *When Prophecy Fails: A Social and Psychological Study of a Modern Group That Predicted the Destruction of the World*. Harper and Row, 1956.
3 Feynman R. The scientific method. YouTube. https://www.youtube.com/watch?v=p2xhb-SdK0g
4 Feynman R. What is science? *Physics Teacher* 1969;67:313–320.
5 Basterfield C, Lilienfeld SO, Bowes SM, Costello TH. The Nobel disease: When intelligence fails to protect against irrationality. *Skeptical Inquirer* 2020;44:4. https://skepticalinquirer.org/2020/05/the-nobel-disease-when-intelligence-fails-to-protect-against-irrationality/
6 Vraga E, Bode L. Defining misinformation and understanding its bounded nature: Using expertise and evidence for describing misinformation. *Political Communication* 2020;37:136–144.
7 Oreskes N. The scientific consensus on climate change. *Science* 2004;306:1886; Cook J, Nuccitelli D, Green SA, et al. Quantifying the consensus on anthropogenic global warming in the scientific literature. *Environmental Research Letters* 2013;8:024024; Powell J. Scientists reach 100% consensus on anthropogenic global warming. *Bulletin of Science, Technology, and Society* 2017;37:183–184; Lynas M, Houlton BZ, Perry S. Greater than 99% consensus on human caused climate change in the peer-reviewed scientific literature. *Environmental Research Letters* 2021;16:114005.

8   Farnsworth SJ, Lichter SR. The structure of scientific opinion on climate change. *International Journal of Public Opinion Research* 2012;24:93–103; Verheggen B, Strengers B, Cook J, et al. Scientists' views about attribution of global warming. *Environmental Science and Technology* 2014;48:8963–8971; Cook J, Oreskes N, Doran PT, et al. Consensus on consensus: A synthesis of consensus estimates on human-caused global warming. *Environmental Research Letters* 2016;11:0148002; Myers KF, Doran PT, Cook J, Kotcher JE, Myers TA. Consensus revisited: Quantifying scientific agreement on climate change and climate expertise among Earth scientists 10 years later. *Environmental Research Letters* 2021;6:104030.
9   IPCC, 2022. *Climate Change 2022: Impacts, Adaptation, and Vulnerability*. Contribution of Working Group II to the Sixth Assessment Report of the Intergovernmental Panel on Climate Change (Pörtner H-O, Roberts DC, Tignor M, et al., eds.). Cambridge University Press, 2022.
10 Pew Research Center. More say there is solid evidence of global warming. Pewresearch.org. October 15, 2012. https://www.pewresearch.org/politics/2012/10/15/more-say-there-is-solid-evidence-of-global-warming/
11 YouGov. The Economist/YouGov Poll. July 17–20, 2021. https://docs.cdn.yougov.com/1aaz80mjhy/econTabReport.pdf
12 Tysen A, Kennedy B. Two-thirds of Americans think government should do more on climate. Pewresearch.org. June 23, 2020. https://www.pewresearch.org/science/2020/06/23/two-thirds-of-americans-think-government-should-do-more-on-climate/
13 Motta M, Chapman D, Stecula D, Haglin K. An experimental examination of measurement disparities in public climate change beliefs. *Climatic Change* 2019;154:37–47.
14 Leiserowitz A, Maibach E, Rosenthal S, et al. Global warming's six Americas, September 2021. Yale University and George Mason University. Yale Program on Climate Change Communication. January 12, 2022. https://climatecommunication.yale.edu/publications/global-warmings-six-americas-september-2021/
15 Egan PJ, Mullin M. Turning personal experience into political attitudes: The effect of local weather on American's perceptions about global warming. *Journal of Politics* 2012;74:796–809.
16 The Associated Press-NORC Center for Public Affairs Research. Where do Americans stand on climate and energy policy? Apnorc.org. October 26, 2021. https://apnorc.org/projects/where-do-americans-stand-on-climate-and-energy-policy/
17 Egan PJ, Mullin M. Recent improvement and projected worsening of weather in the United States. *Nature* 2016;532:357–360.
18 Wright P. 87 percent of Americans unaware there's scientific consensus on climate change. Weather.com. July 11, 2017. https://weather.com/science/environment/news/americans-climate-change-scientific-consensus
19 Nugent C. YouTube has been "actively promoting" videos spreading climate denialism, according to new report. TIME.com. January 16, 2020. https://time.com/5765622/youtube-climate-change-denial/; AVAAZ. Why is YouTube broadcasting climate misinformation to millions? Avaaz report. January 15, 2020. https://avaazimages.avaaz.org/youtube_climate_misinformation_report.pdf
20 Perry MJ. There is no climate emergency say 500 experts in letter to the United Nations. Aei.org. October 1, 2019. https://www.aei.org/carpe-diem/there-is-no-climate-emergency-say-500-experts-in-letter-to-the-united-nations/
21 Benestad RE, Nuccitelli D, Lewandowsky S, et al. Learning from mistakes in climate research. *Theoretical and Applied Climatology* 2016;126:699–703.
22 Holden E. How the oil industry has spent billions to control the climate change conversation. *The Guardian*. January 8, 2020. https://www.theguardian.com/business/2020/jan/08/oil-companies-climate-crisis-pr-spending; Colman Z, Mathiesen K. Climate scientists

take swipe at Exxon Mobil, industry in leaked report. Politico.com. July 2, 2022. https://www.politico.com/news/2021/07/02/climate-scientists-exxon-mobile-report-497805
23. Kennedy B, Tyson A, Funk C. Americans' trust in scientists, other groups declines. Pew Research Center. February 15, 2022. https://www.pewresearch.org/science/2022/02/15/americans-trust-in-scientists-other-groups-declines/
24. Funk C, Hefferon M, Kennedy B, Johnson C. Trust and mistrust in American's views of scientific experts. Pew Research Center. August 2019. https://www.pewresearch.org/science/2019/08/02/trust-and-mistrust-in-americans-views-of-scientific-experts/
25. McCright AM. Anti-reflexivity and climate change skepticism in the US general public. *Society of Human Ecology* 2016;22:77–108.
26. McCright AM, Dunlap RE. Cool dudes: The denial of climate change among conservative white males in the United States. *Global Environmental Change* 2011;21:1163–1172; Bolsen T, Druckman JN. Do partisanship and politicization undermine the impact of a scientific consensus message about climate change? *Group Processes and Intergroup Relations* 2018;21:389–402.
27. Dunlap RE, Jacques PJ. Climate change denial books and conservative think tanks: Exploring the connection. *American Behavioral Scientist* 2013;57:699–731.
28. Brulle RJ. Obstructing action: Foundation funding and US climate change countermovement organizations. *Climatic Change* 2021;166:17.
29. Milman O, Harvey F. US is a hotbed of climate change denial, major global survey finds. *The Guardian.* May 8, 2019. https://www.theguardian.com/environment/2019/may/07/us-hotbed-climate-change-denial-international-poll; YouGov. YouGov Cambridge Globalism Project—Conspiracy Theories. YouGov Poll. August 13–23, 2018. https://d25d2506sfb94s.cloudfront.net/cumulus_uploads/document/2c6lta5kbu/YouGov%20Cambridge%20Globalism%20Project%20-%20Conspiracy%20Theories.pdf
30. Bell J, Poushter J, Fagan M, Huang C. In response to climate change, citizens in advanced economies are willing to alter how they live and work. Pew Research Center. September 14, 2021. https://www.pewresearch.org/global/2021/09/14/in-response-to-climate-change-citizens-in-advanced-economies-are-willing-to-alter-how-they-live-and-work/
31. McCright AM, Dunlap RE. Anti-reflexivity: The American conservative movement's success in undermining climate science and policy. *Theory, Culture, and Society* 2010;27:100–133; McCright AM, Dentzman K, Charters M, Dietz T. The influence of political ideology on trust in science. *Environmental Research Letters* 2013;8:044029.
32. Campbell TH, Kay AC. Solution aversion: On the relation between ideology and motivated disbelief. *Journal of Personality and Social Psychology* 2014;107:809–824.
33. Hatchett F. Man denied kidney transplant at Atrium Health Wake Forest Baptist due to unvaccinated status. WXII12.com. February 8, 2022. https://www.wxii12.com/article/man-denied-kidney-transplant-at-atrium-health-wake-forest-baptist-due-to-unvaccinated-status/39016736#
34. Mark J. He's declining a coronavirus vaccine at the expense of a lifesaving transplant: "I was born free, I'll die free." *The Washington Post.* January 31, 2022. https://www.washingtonpost.com/nation/2022/01/31/chad-carswell-kidney-coronavirus-vaccine/
35. Giordano A. Air Force veteran forced to choose between getting vaccine or dying. *The Epoch Times.* March 18, 2022. https://www.theepochtimes.com/air-force-veteran-forced-to-choose-between-getting-vaccine-or-dying_4345065.html
36. Mandelbaum E. Troubles with Bayesianism: An introduction to the psychological immune system. *Mind & Language* 2019;34:141–157.
37. Hayes SC, Smith S. *Get Out of Your Mind and Into Your Life: The New Acceptance and Commitment Therapy.* New Harbinger Publications, Inc, 2005.
38. Ginges J, Atran S. Psychology out of the laboratory: The challenge of violent extremism. *American Psychologist* 2011:507–519.

39 Strohminger N, Nichols S. The essential moral self. *Cognition* 2014;131:159–171; Heiphetz L, Strohminger N, Young LL. The role of moral beliefs, memories, and preferences in representations of identity. *Cognitive Science* 2017;41:744–767; Heiphetz L, Strohminger N, Gelman SA, Young LL. Who am I? The role of moral beliefs in children's and adults' understanding of identity. *Journal of Experimental Social Psychology* 2018;78:210–219.

40 Strohminger N, Nichols S. Neurodegeneration and identity. *Psychological Science* 2015;26:1469–1479.

41 Graham J, Haidt J, Koleva S, et al. Moral foundations theory: The pragmatic validity of moral pluralism. In: Devine P, Plant A, eds. *Advances in Experimental Social Psychology*, vol 47. Elsevier, 2013.

42 Haidt J, Graham J, Joseph C. Above and below left-right: Ideological narratives and moral foundations. *Psychological Inquiry* 2009;20:110–119; Graham J, Haidt J, Nosek BA. Liberals and conservatives rely on different sets of moral foundations. *Journal of Personality Processes and Social Psychology* 2009;96:1029–1046.

43 Graham J, Nosek BA, Haidt J. The moral stereotypes of liberals and conservatives: Exaggeration of differences across the political spectrum. *PLoS ONE* 7(12):e50092.

44 Haidt J. What makes people vote Republican? *Edge*. September 8, 2008. https://www.edge.org/conversation/jonathan_haidt-what-makes-people-vote-republican.

45 Wehner P. Jonathan Haidt is trying to heal American's divisions. *The Atlantic*. May 24, 2020. https://www.theatlantic.com/ideas/archive/2020/05/jonathan-haidt-pandemic-and-americas-polarization/612025/

46 Curry OS. What's wrong with moral foundations theory, and how to get moral psychology right. *Behavioral Scientist*. March 26, 2019. https://behavioralscientist.org/whats-wrong-with-moral-foundations-theory-and-how-to-get-moral-psychology-right/; Curry OS, Chesters MJ, Van Lissa CJ. Mapping morality with a compass: Testing the theory of "morality-as-cooperation" with a new questionnaire. *Journal of Research in Personality* 2019;78:106–124.

47 Strupp-Levitsky M, Noorbaloochi S, Shipley A, Jost JT. Moral "foundations" as the product of motivated social cognition: Empathy and other psychological underpinnings of ideological divergence in "individualizing" and "binding" concerns. *PLoS ONE* 2020;15(11):e0241144.

48 Bleske-Rechek A, Nelson LA, Baker JP, Remiker MW, Brandt SJ. Evolution and the trolley problem: People save five over one unless the one is young, genetically related, or a romantic partner. *Journal of Social, Evolutionary, and Cultural Psychology* 2010;4:115–127.

49 Hamid S. America without God. *The Atlantic*. April 2021. https://www.theatlantic.com/magazine/archive/2021/04/america-politics-religion/618072/

50 Zmigrod L. A psychology of ideology: Unpacking the psychological structure of ideological thinking. *Perspectives on Psychological Science* 2022 (in press).

51 Pierre JM. Down the conspiracy theory rabbit hole: How does one become a follower of QAnon? In: Miller MK, ed. *The Social Science of QAnon*. Cambridge University Press, 2023, 17–32; Pierre JM. Conspiracy theory belief: A sane response to an insane world? *Review of Philosophy and Psychology* 2023. https://doi.org/10.1007/s13164-023-00716-7

52 Borum R. Radicalization into violent extremism II: A review of conceptual models and empirical research. *Journal of Strategic Security* 2011;4:37–62; Kruglanski AW, Gelfand MJ, Belanger JJ, Sheveland A, Hetiarachchi M, Gunaratna R. The psychology of radicalization and deradicalization: How significance quest impacts violent extremism. *Advances in Political Psychology* 2014;35:69–93; McCauley C, Moskalenko S. Understanding political radicalization: The two-pyramids model. *American Psychologist* 2017;72:205–216; Franks B, Bangerter A, Bauer MW, Hall M, Noort MC. Beyond "monologicality"? Exploring conspiracist worldviews. *Frontiers in Psychology* 2017;8:861.

53 Pierre JM. Conspiracies gone wild: A psychiatric perspective on conspiracy theory belief, mental illness, and the potential for lone actor ideological violence. *Terrorism and Political Violence* 2024. https://doi.org/10.1080/09546553.2024.2329079.
54 Uzarevic F, Coleman TJ, III. The psychology of nonbelievers. *Current Opinion in Psychology* 2021;40:131–138.
55 Hoffer E. *The True Believer: Thoughts on the Nature of Mass Movements*. Harper & Row, 1951.
56 Rip B, Vallerand RJ, Lafrenière M-AK. Passion for a cause, passion for a creed: On ideological passion, identity threat, and extremism. *Journal of Personality* 2012;80:573–602; Webber D, Kruglanski AW. Psychological factors in radicalization: A "3N" approach. In: LaFree G, Freilich JD, eds. *The Handbook of the Criminology of Terrorism*, 1st ed. Wiley, 2017, 479–487; McCauley C, Moskalenko S. Mechanisms of political radicalization: Pathways toward terrorism. *Terrorism and Political Violence* 2008;20:415–433; McCauley C, Moskalenko S. Toward a profile of lone wolf terrorists: What moves an individual from radical opinion to radical action. *Terrorism and Political Violence* 2014;26:69–85; Rip B, Vallerand RJ, Lafrenière M-AK. Passion for a cause, passion for a creed: On ideological passion, identity threat, and extremism. *Journal of Personality* 2012;80:573–602; van den Bos K. Unfairness and radicalization. *Annual Review of Psychology* 2020;71:563–588.
57 Gill P, Farnham F, Clemmow C. The equifinality and multifinality of violent radicalization and mental health. In: Bhui K, Bhugra D, eds. *Terrorism, Violent Radicalization, and Mental Health*. Oxford University Press, 2021, 125–136.
58 Dickson EJ. Former QAnon followers explain what drew them in—and got them out. *Rolling Stone*. September 23, 2020. https://www.rollingstone.com/culture/culture-features/ex-qanon-followers-cult-conspiracy-theory-pizzagate-1064076/
59 Andrews TM. He's a former QAnon believer. He doesn't want to tell his story, but thinks it might help. *The Washington Post*. October 24, 2020. https://www.washingtonpost.com/technology/2020/10/24/qanon-believer-conspiracy-theory/
60 Lord B, Naik R. He went down the QAnon rabbit hole for almost two years. Here's how he got out. CNN.com. October 18, 2020. https://www.cnn.com/2020/10/16/tech/qanon-believer-how-he-got-out/index.html
61 Jadeja J. I left QAnon in 2019. But I'm still not free. Politico.com. December 11, 2021. https://www.politico.com/news/magazine/2021/12/11/q-anon-movement-former-believer-523972
62 Rousselet M, Duretete O, Hardouin JB, Grall-Bronnec M. Cult-membership: What factors contribute to joining or leaving? *Psychiatry Research* 2017;257:27–33.
63 Horgan J. Deradicalization or disengagement? A process in need of clarity and a counterterrorism initiative in need of evaluation. *Revista de Psicología Social* 2009;24:291–298; Rabasa, Pettyjob SL, Ghez JJ, Boucek C. *Deradicalizing Islamic Extremists*. RAND Corporation, 2010.
64 We see things not as they are, but as we are. Quoteinvestigator.com. March 9, 2014. https://quoteinvestigator.com/2014/03/09/as-we-are/

Chapter 10

1 Pennycook G, Fugelsang JA, Koehler DJ. Everyday consequences of analytic thinking. *Current Directions in Psychological Science* 2015;24:425–432; Pennycook G, Cheyne JA, Koehler DJ, Fugelsang JA. On the belief that beliefs should change according to evidence: Implications for conspiratorial, moral, paranormal, political, religious, and science

beliefs. *Judgment and Decision Making* 2020;15:476–498; Bowes SM, Tasimi A. Clarifying the relations between intellectual humility and pseudoscience beliefs, conspiratorial ideation, and susceptibility to fake news. *Journal of Research in Personality* 2022;98:104220; Kwek A, Peh L, Tan J, Lee JX. Distractions, analytical thinking and falling for fake news: A survey of psychological factors. *Humanities and Social Science Communications* 2023;10:319.

2 Leary MP, Diebels KJ, Davisson EK, et al. Cognitive and interpersonal features of intellectual humility. *Personality and Social Psychology Bulletin* 2017;43:793–813; Resnick B. Intellectual humility: The importance of knowing you might be wrong. *Vox.* January 4, 2019. https://www.vox.com/science-and-health/2019/1/4/17989224/intellectual-humility-explained-psychology-replication.

3 Hrishikesh J. What are the chances you're right about everything? An epistemic challenge for modern partisanship. *Politics, Philosophy, & Economics* 2020;19:36–61.

4 Bancroft A. *The Wisdom of the Buddha: Heart Teachings in His Own Words.* Shambhala, 2000.

5 Van de Calseyde PPFM, Efendić E. Taking on a disagreeing perspective improves the accuracy of people's quantitative estimates. *Psychological Science* 2022;33:971–983.

6 Frederick S. Cognitive reflection and decision making. *Journal of Economic Perspectives* 2005;19:25–42.

7 Stanovich KE, West RF. On the relative independence of thinking biases and cognitive ability. *Personality Processes and Individual Differences* 2008;94:672–695.

8 Young DG, Maloney EK, Bleakley A, Langbaum JB. "I feel it in my gut": Epistemic motivations, political beliefs, and misperceptions of COVID-19 and the 2020 presidential election. *Journal of Social and Political Psychology* 2022;10:643–656.

9 Pennycook G, Cheyne JA, Seli P, Koehler DJ, Fugelsang JA. Analytic cognitive style predicts religious and paranormal belief. *Cognition* 2012;123:335–346.

10 Pennycook G, Cheyne JA, Koehler DJ, Fugelsang JA. On the belief that beliefs should change according to evidence: Implications for conspiratorial, moral, paranormal, political, religious, and science beliefs. *Judgment and Decision Making* 2020;15:476–498.

11 Scheffer M, van de Leemput I, Weinans E, Bollen J. The rise and fall of rationality in language. *PNAS* 2021;118:e2107848118.

12 Wageningen University. "We conclude" or "I believe?" Study finds rationality declined decades ago. Phys.org. January 12, 2022. https://phys.org/news/2022-01-rationality-declined-decades.html

13 Vary AB. "Superman changes motto to truth, justice and a better tomorrow," says DC chief. *Variety.* October 16, 2021. https://variety.com/2021/film/news/superman-new-motto-dc-fandome-1235090712/

14 Innocenzi RL, Brown K, Liggit P, Tout S, Tanner A. "Think before you click. Post. Type." Lessons learned from our University Cyber Security Awareness Campaign. *Journal of Cybersecurity Education, Research and Practice* 2018;1:3.

15 Mackintosh E. Finland is winning the war on fake news. What it's learned may be crucial to Western democracy. CNN.com. May 2019. https://edition.cnn.com/interactive/2019/05/europe/finland-fake-news-intl/; Gross J. How Finland is teaching a generation to spot misinformation. *The New York Times.* January 10, 2023. https://www.nytimes.com/2023/01/10/world/europe/finland-misinformation-classes.html

16 Horn S, Veermans K. Critical thinking efficacy and transfer skills defend against "fake news" at an international school in Finland. *Journal of Research in International Education* 2019;18:23–41.

17 Compton J, van der Linden S, Cook J, Basol M. Inoculation theory in the post-truth era: Extant findings and new frontiers for contested science, misinformation, and conspiracy theories. *Social and Personality Psychology Compass* 2021;15: e12602; Lewandowsky

S, van der Linden S. 2021. Countering misinformation and fake news through inoculation and prebunking. *European Review of Social Psychology* 32:348–384; van der Linden S. Misinformation: Susceptibility, spread, and interventions to immunize the public. *Nature Medicine* 2022;28:460–467; Lu C, Hu B, Li Q, Bi C, Ju X-D. Psychological inoculation for credibility assessment, sharing intention, and discernment of misinformation: Systematic review and meta-analysis. *Journal of Medical Internet Research* 2023;25:e49255.

18 Nyhan B, Reifler J. When corrections fail: The persistence of political misperceptions. *Political Behavior* 2020;32:303–330; Wood T, Porter E. The elusive backfire effect: Mass attitudes' steadfast factual adherence. *Political Behavior* 2019;41:135–163.

19 Salter J. Judge limits Biden administration in working with social media companies. APnews.com. July 4, 2023. https://apnews.com/article/social-media-protected-speech-lawsuit-injunction-148c1cd43f88a0284d5a3c53fd333727

20 Kosseff J. America's favorite flimsy pretext for limiting free speech. *The Atlantic*. January 4, 2022. https://www.theatlantic.com/ideas/archive/2022/01/shouting-fire-crowded-theater-speech-regulation/621151/

21 Vogels EA, Anderson M, Porteus M, et al. Americans and "cancel culture": Where some see calls for accountability, others see censorship, punishment. Pew Research Center. May 19, 2021. https://www.pewresearch.org/internet/2021/05/19/americans-and-cancel-culture-where-some-see-calls-for-accountability-others-see-censorship-punishment/

22 The Editorial Board. America has a free speech problem. *The New York Times*. March 18, 2022. https://www.nytimes.com/2022/03/18/opinion/cancel-culture-free-speech-poll.html

23 St. Aubin C, Liedke J. Most Americans favor restrictions on false information, violent content online. Pew Research Center. July 20, 2023. https://www.pewresearch.org/short-reads/2023/07/20/most-americans-favor-restrictions-on-false-information-violent-content-online/

24 Kozyreva A, Herzog SM, Lewandowsky S, et al. Resolving content moderation dilemmas between free speech and harmful misinformation. *PNAS* 2023;120:e2210666120.

25 Linder NM, Nosek BA. Alienable speech: Ideological variations in the application of free-speech principles. *Political Psychology* 2009;30:67–92.

26 Shellenberger M. The new war on free speech. Unherd.com. June 19, 2023. https://unherd.com/2023/06/the-new-world-war-on-free-speech/

27 Lakier G. The great free-speech reversal. *The Atlantic*. January 27, 2021. https://www.theatlantic.com/ideas/archive/2021/01/first-amendment-regulation/617827/

28 Hammond-Errey M. Elon Musk's Twitter is becoming a sewer of disinformation. Foreignpolicy.com. July 15, 2023. https://foreignpolicy.com/2023/07/15/elon-musk-twitter-blue-checks-verification-disinformation-propaganda-russia-china-trust-safety/

29 Zilber A. Elon Musk's X to hire 100 content moderators in wake of Taylor Swift, AI fiasco. *The New York Post*. January 29, 2024. https://nypost.com/2024/01/29/business/elon-musks-x-to-hire-100-content-moderators-in-wake-of-taylor-swift-ai-fiasco/

30 Pretus C, Javeed AM, Hughes D, et al. The *Misleading* count: An identity-based intervention to counter partisan misinformation sharing. *Philosophical Transactions B* 2023;379:20230040.

31 Geisel TS. *The Lorax*. Random House, 1971.

Index

For the benefit of digital users, indexed terms that span two pages (e.g., 52–53) may, on occasion, appear on only one of those pages.

absence of evidence, 10, 157–58
acceptance and commitment therapy (ACT), xiii, 166
acrimonious argumentation, 37–38
activists, 174–75
affective polarization, 136–41
alternative facts, 80–85, 127
American Conspiracy Theories (Uscinski), 89–90
America's Frontline Doctors (AFD), 105
analytical thinking, 10, 12, 61, 94–95, 97, 115, 121, 122, 130, 131, 150, 172, 178, 180–88
anecdotal bias, 8
anthropogenic climate change (ACC), 160–61, 162–65. *See also* climate change denial
anti-intellectualism, 59, 101
anti-vaccine beliefs, 45, 52–53, 72–75
apostates, 175–76
Arendt, Hannah, 83
artificial intelligence (AI), 28, 35, 76
Asimov, Isaac, 59, 101
Austin, Melanie, 36

base rate fallacy, 15, 16–17
beliefs/believers
   activists, 174–75
   anti-vaccine beliefs, 45, 52–53, 72–75
   apostates, 175–76
   climate change denial and, 106, 156, 158–65, 166, 185, 188
   climate science, 162–65
   conspiracy belief, 93, 101
   conspiracy theory convictions, 88–89
   defined, 7
   delusion-like, xi, 5, 6–7, 12, 17, 33–34, 62–63, 69, 88, 96, 111, 181, 188
   factual, 23, 156, 165–67, 169–70, 172, 176, 179–80
   fence-sitters, 172–73
   flat-Earth believers, 86–89
   ideological commitment and, 170–76
   misbeliefs, ix–x, 7–12, 17, 22–23
   moral absolutism and, 167–70, 178
   moral relativism and, 167–70, 172, 174
   naive realism and, 9, 14, 16–17, 27, 29–30, 37, 44–46, 65, 87–88, 159–62
   nonbelievers and, 171, 172
   personal, 13–14, 36, 41, 181–82
   in personal control over circumstances, 18–20
   as probability judgments, 13–17
   religious, 42–46
   seeing is believing, 9, 87, 162
   shared, 3–4, 33–34, 42–43, 97
   true believers, 173–74
   truth seekers, 155–58
   See also divisiveness over beliefs; false beliefs
better-than-the-average effect, 17–18
bias
   anecdotal, 8
   cognitive, xii, 4, 16–18, 25, 29–30, 35, 36–37, 59, 94, 114, 142–43, 151, 179
   disconfirmation, 58, 137–38
   implicit, 142–46
   myside, 137–38, 163–64, 189
   See also confirmation bias
bias blind spot, 29–30, 114–15, 134–35. *See also* confirmation bias
biased assimilation, 29
Biden, Joe, 53, 64, 67–68, 82, 90–91, 135–36, 184
Bierce, Ambrose, 20
biofields, 117, 119–20, 121
Black Lives Matter movement, 79–80, 146–47
Blair, Tony, 83
Bollinger, Charlene, 74–75
Bollinger, Ty, 74–75
bots (automated computer programs) use, 75–76
Brandolini, Alberto, 122
Brexit, 90–91
Brogan, Kelly, 72–75
Brontiakowski, David, 76
Brown, Jonathan, 17
Buddha, 27
bullshit
   defined, 131–32
   detectors, 60–61
   evasive bullshitting, 126–28, 129–30
   fake news and, 113–14, 115, 131
   false beliefs and, 121
   introduction to, 111
   malingering, 128–32
   overview of, 112–15
   politics of, 126–28
   postmodern premises of, 122–26
   pseudoscience and, 116–22

bullshit blind spot, 114–15, 125, 128
Bullshit Receptivity Scale (BRS), 114
Bullshitting Frequency Scale (BFS), 114–15
Bush, George H. W., 83
Byrne, John, 11–12

Carlson, Tucker, 56–59
Carswell, Chad, 165–70
censureship and censorship, 183–86
Center for Countering Digital Hate (CCDH), 74–75
Centers for Disease Control and Prevention (CDC), 31–32, 45, 99–100
Ceselkoski, Mirko, 76
ChatGPT, 76, 183
Cheek defense, 62–64
Chopra, Deepak, 116–18, 119–20, 124
circular inference, 8
citizen journalism, 50
clickbait, 51–52, 76
climate change denial, 90–91, 106, 156, 158–65, 166, 185, 188
climate science, 162–65
Clinton, Bill, 83
Clinton, Hillary, 147
Coast to Coast AM radio show, 46
cognitive behavioral therapy (CBT), 5–7, 10, 177
cognitive biases, xii, 4, 16–18, 25, 29–30, 35, 36–37, 59, 94, 114, 142–43, 151, 179
cognitive dissonance, 146, 156–58, 164–66, 171–72, 173, 175–76, 179
cognitive distortions, 2, 5–8, 11, 13–14, 98, 177
cognitive flexibility, 166, 172, 178, 179–80
cognitive laziness, 59–61
cognitive processes, ix, 24, 57, 65–66
Cognitive Reflection Test (CRT), 180–81
Computational Propaganda Research Project, 80
confirmation bias
  biased assimilation, 29
  defined, 29
  doing one's own research, 101–3
  false beliefs and, 27, 41
  filter bubble and, 35–36, 41
  Heaven's Gate cult and, 42–44
  myside bias, 137–38, 163–64, 189
  online impact on, 30–41
  overview of, 29–30
  peripheral brains in, 27–29
conflict resolution, 188–90
conjunction fallacy, 94–95
conspiracist ideation, 93, 97, 100–1, 134–35
conspiracy belief, 93, 101
conspiracy mentality, 89, 93, 94–95, 100, 103, 111
conspiracy theories
  belief conviction in, 88–89

COVID-19 pandemic and, 45, 52–53, 90, 91–92, 94, 99, 104–6, 119, 165–70, 173
  doing one's own research, 101–3
  expressing allegiance with, 106–10
  false beliefs and, 95, 110
  introduction to, xi
  mass psychosis and, 97, 139
  misbeliefs and, 96–106
  misinformation and, 96–106
  overview of, 89–92
  post-truth world and, 86–89
  psychology of, 93–95
  QAnon and, 44, 89, 90–92, 106–9, 121, 133–34, 157, 175–76, 184, 188
  trust and, 98–101
Conway, Kellyanne, 81–82
Council for National Policy, 105
COVID-19 pandemic, 45, 52–53, 72–75, 76, 119
COVID-19 pandemic conspiracy theories, 45, 52–53, 90, 104–6, 119, 121, 165–70, 173
cult of ignorance, 59

The Death of Expertise: The Campaign Against Knowledge and Why It Matters (Nichols), 84–85
deliberative reasoning, 16
delusional parasitosis, 28
delusion-like beliefs, xi, 5, 6–7, 12, 17, 33–34, 62–63, 69, 88, 96, 111, 181, 188
delusions
  defined, 2, 3–5
  false beliefs and, 7
  fixed, false beliefs and, 2, 3–4
  gang-stalking, 32–34
  introduction to, xi
  online impact on, 30–34
  overview of, 1–5
  paranoid, 8, 32, 98
  pathological, xii, 7–8, 109
democratic backsliding, 140
Dennett, Daniel, 7, 11–12, 125
Derrida, Jacques, 122–23
Devil's Dictionary (Bierce), 20
disconfirmation bias, 58, 137–38
disinformation ecosystem, 67
disinformation industrial complex
  alternative facts, 80–85
  anti-vaccine disinformation, 45, 72–75
  apex predators of, 67–76
  bots (automated computer programs) use, 75–76
  distributed responsibility/liability, 62–64
  false beliefs and, 63, 64–65, 68, 69, 84–85
  just-asking-questions, 70–72, 75–76, 102
  misinformation and, 65–67
  mistrust and, 65–67

political propaganda, 77–80
truth effect and, 77–80
war on information, 67–70
distributed responsibility/liability, 62–64
divisiveness over beliefs
    affective polarization, 136–41
    fake news and, 134–35
    false beliefs and, 134–35
    identity threat, 146–49
    implicit bias and, 142–46
    in marriages, 133–34, 152–54
    political polarization, xii, 39, 58–59, 60–61, 135–37, 140, 142, 149–53, 168
    in politics, 134–41
    race politics, 141–49
    sectarianism, 138–41
    tribalism and, 142, 146–47, 151
doing one's own research, 101–3
Dominion Voting Systems, 85
Doughty, Terry, 184
Drummond, Margo, 62–64
Drummond, Robert, 62–64, 111, 114–15
Dunbar, Frank, 30–31, 33
Dunning, David, 23–26, 113–14
Dunning-Kruger effect, 23–26, 113–14, 134–35, 157–58, 162–63, 178

education reform, 182–83
Elliott, Jane, 141
epistemic authority, 65–67, 69, 72, 80, 84–85, 100, 101, 187–88
epistemic mistrust, 65–67, 98–99, 100, 101, 103–4, 115
epistemic vigilance, 65–66, 98
errors of belief, 5–6. *See also* cognitive distortions
Evans, Anthony, 118
evasive bullshitting, 126–28, 129–30

fact-checking, 54, 59, 85, 128, 130, 131, 184
factual beliefs, 23, 156, 165–67, 169–70, 172, 176, 179–80
Fairness Doctrine (1949), 51
faith, 10, 11–12
faith in intuition, 94–95, 115, 119, 180, 181
fake news
    bullshit and, 113–14, 115, 131
    divisiveness over beliefs, 134–35
    false news and, 52, 56–57, 60, 65, 66–67, 79–80, 132
    introduction to, ix–x
    truth effect and, 77–80
    See also misinformation
false beliefs
    bullshit and, 121
    confirmation bias and, 27, 41
    conspiracy theories and, 95, 110

delusions and, 7
disinformation industrial complex and, 63, 64–65, 68, 69, 84–85
divisiveness over, 134–35
fixed, 2, 3–4
introduction to, ix–x, xi–xii, xiii
misinformation and, 42–44, 45–46, 52–53, 61
overconfidence and, 16–17
See also delusions
false memories, 21–23
false news, 52, 56–57, 60, 65, 66–67, 79–80, 132
Fauci, Anthony, 99–100
Feather, William, 24–25
fence-sitters, 172–73
Festinger, Leon, 156–57
Feynman, Richard, 158–59
filter bubble, 35–36, 41
Fisher, Matthew, 27
flat-Earth believers, 86–89
Frankfurt, Harry, 112, 114
Freud, Sigmund, xii

gambler's fallacy, 13–17
gang-stalking, 32–34
Gawande, Atul, 10, 158–59
Geiderman, Joel, 11–12
Geim, Andre, 113–14
Goebbels, Joseph, 77
Goldilocksian ideal, 17, 18, 19–21, 25
Goop, Inc, 70–73, 119
Gore, Al, 101
The Greatest Hoax: How the Global Warming Conspiracy Threatens Your Future (Inhofe), 106
Griffin, Dale, 14–15
Griffin, John Howard, 154
Grizzly Man (2005), 19, 20–21
gut instinct, 8–9, 65, 96

Haidt, Jonathan, 167–68
hallucinations, xi, 3, 9
Hamid, Shahi, 170–71
harmful misinformation, 53, 158–59, 184, 187
healthy lies, 17–21
Heaven's Gate, 42–44
Henley, William Ernest, 19
Herzog, Werner, 19, 20–21
heuristics, 16–17, 60
Hitler, Adolf, 77
Hobart, Byrne, 86
Hughes, Scottie Nell, 81
human brain, ix, 16, 55

identity politics, 134–41
identity protection, 56–59
identity threat, 146–49

ideological commitment, 170–76
ideological passion, 11, 174–75
Ig Nobel Prize, 113–14, 117–18, 122
illusion of objectivity, 59
Implicit Association Test (IAT), 142–46
implicit bias, 142–46
InfoWars, 44, 68–69, 71, 103–4, 175
Inhofe, James, 106, 161–62
intellectual humility, 178–79
Internet Research Agency (IRA), 79–80
intuition, 8–9, 10, 13–16, 27, 29, 57, 60, 65, 94–95, 115, 119–20, 130, 172, 180–82
intuitive cognitive style, 115
intuitive thinking, 150, 180, 181–82
Invictus (Henley), 19
Irving, Kyrie, 44–45, 86

Jolley, Daniel, 91
Jones, Alex, 68–69
Jost, John, 144–45
jumping to conclusions, 5–6, 8, 16
just-asking-questions, 70–72, 75–76, 102

Kahan, Dan, 58, 60
Kahneman, Daniel, 14, 15, 57, 168, 180
Kavanagh, Jennifer, 50–51
Koppel, Ted, 47–49, 51, 56–57, 59
Kruger, Justin, 23–26, 113–14
Kubin, Emily, 153–54
Kunda, Ziva, 59
Kuyper, Abraham, 170–71

Lacan, Jacques, 123
leap of faith, 10
lesbian, gay, bisexual, transgender, and queer (LGBTQ) rights, 79–80
Littrell, Shane, 127–28
locus of control, 18–20
Loftus, Elizabeth, 21–23

Mackay, Charles, xii–xiii
Maddow, Rachel, 56–59
malingering, 128–32, 133–34, 152–54
mass psychosis, 97, 139
McCain, John, 135–36
McCright, Aaron, 163–64
McCrummen, Stephanie, 36
McKay, Ryan, 7
medical misinformation, 45, 52–53
Mein Kampf (Hitler), 77
mental illness, x–xiii. *See also* schizophrenia
Mercola, Joseph, 75
misbeliefs, ix–x, 7–12, 17, 22–23, 96–106
misinformation
    censureship and censorship, 183–86
    cognitive laziness and, 59–61

conspiracy theories and, 96–106
dangers of, 42–46
disinformation industrial complex and, 65–67
fake news and, 50–55
false beliefs and, 42–44, 45–46, 52–53, 61
false news, 52, 56–57, 60, 65, 66–67, 79–80, 132
harmful, 53, 158–59, 184, 187
Heaven's Gate and, 42–44
Hughes, "Mad Mike" and, 44–45
identity protection and, 56–59
medical misinformation, 45, 52–53
motivated reasoning in, 56–59
Pizzagate and, 44
political TV news and, 46–50
profitability of, 52–53
See also fake news
mistrust, 65–67, 98–99, 100, 101, 103–4, 115. *See also* epistemic mistrust
Montaigne, Michel del, 11
moral absolutism, 167–70, 178
moral outrage, 37–38
moral relativism, 167–70, 172, 174
motivated disbelief, 58, 137–38
motivated reasoning, 56–59
motivated skepticism, 58
Murdoch, Rupert, 67–68
myside bias, 137–38, 163–64, 189
The Myth of Repressed Memory (Loftus), 21–23

naive realism, 9, 14, 16–17, 27, 29–30, 37, 44–46, 65, 87–88, 159–62
National Aeronautics and Space Administration (NASA), 87–88
New Age Movement, 116
Nguyen, C. Thi, 36–37
Nichols, Tom, 84–85
1984 (Orwell), 79
nonbelievers, 171, 172

Obama, Barack, 36, 77–78, 81, 83, 95, 147
Obama "birther" conspiracy, 89, 95, 104, 107–8
objective evidence, 5–6, 8–9, 10–12, 29, 32–33, 43, 57, 65, 94–95, 99, 113, 158–60, 162–63, 172, 175–76, 177–78, 179–80, 186
On Bullshit (Frankfurt), 112
One America News Network (OANN), 67–68
online disinhibition effect, 39
open-access publishing, 54–55, 72–73
optimism, 17, 20–21, 124, 191
O'Reilly, Bill, 47–48
The O'Reilly Factor TV show, 47–48
Orwell, George, 79
overconfidence
    beliefs as probability judgments, 13–17
    better-than-the-average effect, 17–18
    Dunning-Kruger effect and, 23–26

false beliefs and, 16–17
false memories, 21–23
healthy lies, 17–21
personal control over circumstances, 18–20
positive illusions and, xii, 17, 18–19, 20–21, 25, 165–66

Paltrow, Gwenyth, 70–72, 119, 124
paranoia, xi, 1, 3, 8, 32, 65–66, 95–98, 100–2, 103–4, 109–10, 115, 181
paranoid delusions, 8, 32, 98
paranormal, ix–x, 46, 101–2, 115, 181
Pariser, Eli, 35, 36
pathological delusion, xii, 7–8, 109
Payne, B. Keith, 144–45
Pennycook, Gordon, 113–15, 117–18, 119, 122, 130–31
perception is reality, 9, 87
peripheral brains, 27–29, 39, 41, 183
Perkins, Cecily, 33–34
personal beliefs, 13–14, 36, 41, 181–82
personal control over circumstances, 18–20
personal truth, 9, 182
Petrocelli, John, 113
Piff, Paul, 19–20
Pirsig, Robert, 11, 171–72
Plato, 3
police brutality, 79–80
political divisiveness, 134–41
political polarization, xii, 39, 58–59, 60–61, 135–37, 140, 142, 149–52, 168
political propaganda, 77–80
political sectarianism, 138–41
political TV news coverage, 46–50
Pomerantsev, Peter, 80–81
positive illusions, xii–xiii, 17, 18–19, 20–21, 25, 165–66
Post, Neil, 111
post-truth world
    academic analysis of, 84
    alternative facts and, 80–85
    analytical thinking and, 10, 12, 61, 94–95, 97, 115, 121, 122, 130, 131, 150, 172, 178, 180–88
    bullshit and, 113, 127
    censureship and censorship, 183–86
    climate change denialism, 159–60
    cognitive flexibility and, 166, 172, 178, 179–80
    conflict resolution, 188–90
    conspiracy theories and, 86–89, 101
    diagnosis and cure, 177–78
    education reform, 182–83
    intellectual humility, 178–79
    introduction to, ix–x
    intuitive thinking, 150, 180, 181–82
    truth detection, 178–82
    vox populi, vox dei phrase, 186–88

predatory publishers, 54–55
preprints, 55
probability judgments, 13–17
pseudoscience, xii, 116–22
psychiatric disorders, x–xiii, 5, 6–7, 19, 72–73
psychopathology, xi–xii, xiii, 98–99, 109, 150–51, 171
The Psychopathology of Everyday Life (Freud), xii
public opinion, 79–80, 104, 116
Putin, Vladimir, 82, 99

QAnon, 44, 89, 90–92, 106–9, 121, 133–34, 157, 175–76, 184, 188

race politics, 141–49
religious beliefs, 42–46
reverse gambler's fallacy, 15
Rich, Michael, 50–51
Richeson, Jennifer, 148
Riess, Adam, 155–56, 159, 176
Roberts, Vernon, 1–2, 33
Rosa, Emily, 120

Sagan, Carl, 59
Sanchez, Carmen, 24
Sandy Hook Elementary School shooting, 68, 85, 91–92, 138
schizophrenia, xi, 1, 2, 3, 5–7, 8, 30–31, 33, 109, 118, 155–56
Schnitzler, Arthur, 155
Scientific Bullshit Receptivity Scale (SBRS), 118–19
Scott, Walter, 62
sectarianism, 138–41
seeing is believing, 9, 87, 162
Seifert, Colleen, 65
self-enhancement, 18
shared beliefs, 3–4, 33–34, 42–43, 97
shared psychotic disorder, 33–34, 42–43
Skeptical Medicine (Byrne), 11–12
social distance, 136–37
social groups, 58, 59, 98–99, 174
Sokal, Alan, 122–27
sovereign citizens, 62–64
Sperber, Dan, 117–18
Spicer, Sean, 81–82
Stamp, Nikki, 72
Stern, Ken, 154
subjective evidence, 8–9
subjective experience, 8, 9, 16, 22–23, 27, 33–34, 43, 65–66, 98
superstitions, ix–x
Swift, Jonathan, 42

talk therapy, x–xi, xiii
Taylor, Shelley, 17–18
Therapeutic Touch (TT), 119–21

This Is Not Propaganda: Adventures in the War Against Reality (Pomerantsev), 80–81
Treadwell, Timothy, 19, 20–21
tribalism, 142, 146–47, 151
trolley problem, 169–70, 179, 186
true believers, 173–74
Trump, Donald
   alternative facts and, 80–85
   Clinton, Hillary and, 147
   COVID-19 conspiracy, 104
   disinformation by, 67–68, 69, 81–83
   divisiveness over, 133–34
   election loss, 95, 157
   election win, 147–48
   interviewing of, 47–48
   Obama "birther" conspiracy, 104
trust and conspiracy theories, 98–101
truth decay, 50–55, 56–57, 66–67, 101
truth detection, 178–82
truth effect, 77–80
Tuskegee Syphilis Study (1932-1979), 99
Tutu, Desmond, 153
Tversky, Amos, 14–15

UFOs, 46, 91
US Capitol riots (2021), 106, 140–41
Uscinski, Joseph, 89–90
US Surgeon General, 99–100

vaccine conspiracy theories, 95, 104–5, 165
vaccine hesitancy, 45, 52–53, 90, 165–70, 173
Veterans Healthcare Administration (VHA), 82
vox populi, vox dei phrase, 186–88

war on information, 67–70
Waterman, Bob, 133
The Weekly World News tabloid, 46
Wiesel, Elie, 153
World Health Organization, 99–100

Yamamoto, Mayu, 113–14
yellow journalism, 50–51

Zen and the Art of Motorcycle Maintenance (Pirsig), 11
Zmigrod, Leor, 171